P9-BYW-159

NUMBER THEORY

CONTEMPORARY UNDERGRADUATE MATHEMATICS SERIES
Robert J. Wisner, Editor

MATHEMATICS FOR THE LIBERAL ARTS STUDENT
Fred Richman, Carol Walker, and Robert J. Wisner

INTERMEDIATE ALGEBRA
Edward D. Gaughan

ALGEBRA: A PRECALCULUS COURSE
James E. Hall

MODERN MATHEMATICS: AN ELEMENTARY APPROACH, SECOND EDITION
Ruric E. Wheeler

FUNDAMENTAL COLLEGE MATHEMATICS: NUMBER SYSTEMS AND INTUITIVE GEOMETRY
Ruric E. Wheeler

MODERN MATHEMATICS FOR BUSINESS STUDENTS
Ruric E. Wheeler and W. D. Peeples

ANALYTIC GEOMETRY
James E. Hall

INTRODUCTORY GEOMETRY: AN INFORMAL APPROACH
James R. Smart

AN INTUITIVE APPROACH TO ELEMENTARY GEOMETRY
Beauregard Stubblefield

GEOMETRY FOR TEACHERS
Paul B. Johnson and Carol H. Kipps

LINEAR ALGEBRA
James E. Scroggs

AN INTRODUCTION TO ABSTRACT ALGEBRA
A. Richard Mitchell and Roger W. Mitchell

INTRODUCTION TO ANALYSIS
Edward D. Gaughan

A PRIMER OF COMPLEX VARIABLES WITH AN INTRODUCTION TO ADVANCED TECHNIQUES
Hugh J. Hamilton

CALCULUS OF SEVERAL VARIABLES
E. K. McLachlan

PROBABILITY
Donald R. Barr and Peter W. Zehna

THEORY AND EXAMPLES OF POINT-SET TOPOLOGY
John Greever

AN INTRODUCTION TO ALGEBRAIC TOPOLOGY
John W. Keesee

NUMBER THEORY: AN INTRODUCTION TO ALGEBRA
Fred Richman

NUMBER THEORY

AN INTRODUCTION TO ALGEBRA

FRED RICHMAN

New Mexico State University

BROOKS/COLE PUBLISHING COMPANY

Belmont, California

A Division of Wadsworth Publishing Co., Inc.

L.C. Cat. Card No: 71–146973
ISBN 0–8185–0006–9

Printed in the United States of America

1 2 3 4 5 6 7 8 9 10—75 74 73 72 71

PREFACE

The basic ideas of modern algebra were forged on the anvil of number theory. It is my opinion that number theory still serves as the best motivation for these ideas and provides the most accessible domain for their application. Accordingly, the first four or five chapters of this text have been used as a one-semester course to provide students with their first brush with abstract algebra—groups, rings, and fields. The idea is to introduce algebraic concepts that will help the student understand specific numerical situations rather than to pick abstract definitions and theorems out of the air and follow them with a few examples.

On the other hand, algebraic notions deepen our understanding of numbers by isolating significant properties and placing them in more general settings. Theorems about numbers that appear slightly artificial and proofs that strike one as a bit ad hoc can often be recast in a broader language that makes both statement and justification simpler and more attractive. By taking the more general point of view, one can avoid the impression that number theory is merely an appendage to the central mathematics curriculum. Both modern algebra and number theory are thereby enriched and made more palatable.

Chapters I through IV deal with the algebraic properties of the ring of integers Z and its residue class rings Z_n. The integers provide an example of a principal ideal domain and a unique factorization domain, while the rings Z_n supply rings with zero-divisors and fields. The groups of units of the rings Z_n form a readily available collection of abelian groups whose structure is simple but not obvious. Attempts to understand these groups lead to discussions of the notions of isomorphism and direct sum, the most important tools in the study of the structure of algebraic systems. Elementary properties of rings of polynomials over a field are developed in order to show that the group of nonzero elements of a finite field is cyclic. This provides the opening wedge for the treatment of primitive roots, thus completing the theory of the groups of units of Z_n.

In Chapter V the problem of representing numbers as sums of two squares leads to the study of the ring of Gaussian integers and draws attention to quadratic residues. An exposition of the reciprocity law is provided in Chapter VI.

Chapters VII, VIII, and IX develop the theory of quadratic fields. Here the student is exposed to a variety of fields and to integral domains where unique factorization fails. Some rudimentary facts about finite-dimensional vector spaces are introduced and the theory of polynomial rings over a field is developed further. The student will also encounter algebraic numbers and integers, prime and maximal ideals, rings modulo ideals, invertible ideals, and a taste of Diophantine equations.

Chapter X provides an application of the elementary algebraic number theory (developed in Chapter VII) to the classical ruler and compass construction problems of Euclidean geometry.

I am indebted to Jeanne Agnew of Oklahoma State University, William E. Briggs of the University of Colorado, J. Richard Byrne of Portland State University, Paul J. McCarthy of the University of Kansas, Theodore Tracewell of California State College at Hayward, and to series editor Robert J. Wisner of New Mexico State University for their helpful reviews of the manuscript.

Fred Richman

CONTENTS

III DECOMPOSITIONS

IV THE STRUCTURE OF Z_n AND U_n

V SUMS OF TWO SQUARES

VI QUADRATIC RESIDUES

VII ALGEBRAIC NUMBER FIELDS

VIII RINGS OF ALGEBRAIC INTEGERS

IX IDEAL THEORY

X RULER AND COMPASS CONSTRUCTIONS

NUMBER THEORY

CHAPTER 0

TERMINOLOGY

We shall be working with sets. A *set* is just what you think it is: a collection of things. We abbreviate the statement "a is in the set A" by writing "$a \in A$." Thus, if $\mathscr{R}$ denotes the set of all real numbers, we can write "$\sqrt{2} \in \mathscr{R}$" and "$\pi \in \mathscr{R}$" instead of "$\sqrt{2}$ is a real number" and "π is a real number." The notation $a, b, c \in A$ means $a \in A$ and $b \in A$ and $c \in A$. The notation $a \notin A$ means a is *not* in A.

If $a \in A$, we say that a is an *element* or a *member* of A. A set can be described by writing down a condition that tells what its elements are—for example, "the set of all real numbers greater than 2" or "the set of whole numbers divisible by 17." We use the notation $\{x \in S \mid P(x)\}$ to indicate the set of all elements in S with the property P; for example,

$\{x \in \mathscr{R} \mid x > 2\}$ is the set of real numbers that are greater than 2. Another way of describing a set is simply to list all its elements. If we wish to consider the set whose elements are the numbers 1, 2, 3, and 4, we write it as $\{1, 2, 3, 4\}$. This set can also be described as the set of positive whole numbers that are less than 5. Thus, there are many ways to describe the same set.

Consider the set of U.S. Senators who have two heads. Compare this to the set of real numbers whose squares are -1. These are two descriptions of the same set. The only thing that matters about a set is which things are in it and which are not, and these two radically different descriptions agree on this point: nothing is in the set. The set that has no elements is called the *empty set* and is denoted by $\varnothing$. The convenience of the concept of the empty set far outweighs any qualms we might have about a set with nothing in it. For example, consider the set of whole numbers n greater than 2 for which there exist positive whole numbers a, b, and c such that $a^n + b^n = c^n$. No one knows whether or not there are any such numbers n, and the determination of this set is the subject of much research. Should we forego referring to it as a set until we find out?

If A and B are sets with the property that every element of A is also an element of B, we say that A is a *subset* of B and write $A \subseteq B$. If $A \subseteq B$ and $A \neq B$, we say that A is a *proper subset* of B. If B is any set, we always have $B \subseteq B$ and $\varnothing \subseteq B$. Why? The subsets of $\{1, 2, 3, 4\}$ are

$$\varnothing, \{1\}, \{2\}, \{3\}, \{4\}, \{1, 2\}, \{1, 3\}, \{1, 4\}, \{2, 3\}, \{2, 4\}, \{3, 4\}, \{1, 2, 3\}, \\ \{1, 2, 4\}, \{1, 3, 4\}, \{2, 3, 4\}, \text{ and } \{1, 2, 3, 4\}.$$

If A and B are sets, we may form a third set, which we write $A \cap B$, consisting of all those elements that are in both A and B—that is,

$$A \cap B = \{x \mid x \in A \text{ and } x \in B\}.$$

Thus, if A is the set of squares of whole numbers and B is the set of numbers less than 5π, then $A \cap B = \{1, 4, 9\}$. Also

$$\{1, 2, 3, 4\} \cap \{2, 3, 5, 7, 11\} = \{2, 3\},$$

whereas $\{2, 4, 6, 8\} \cap \{1, 3\} = \varnothing$. The set $A \cap B$ is called the *intersection* of A and B. If $A \cap B = \varnothing$, we say that A and B are *disjoint*.

Another set formed from the sets A and B is the collection of all elements that are either in A or in B (or in both)—that is, $\{x \mid x \in A$ or

$x \in B\}$. This set is written $A \cup B$ and is called the *union* of A and B. Thus,

$$\{1, 2, 3, 4\} \cup \{2, 3, 5, 7, 11\} = \{1, 2, 3, 4, 5, 7, 11\}$$

and

$$\{2, 4, 6, 8\} \cup \{1, 3\} = \{1, 2, 3, 4, 6, 8\}.$$

Both the notions of intersection and union may be extended in a perfectly natural way to more than two sets. Thus, $A_1 \cap A_2 \cap A_3 \cap A_4$ is the set of elements that are in *all* the sets A_1, A_2, A_3, and A_4, whereas $a \in A_1 \cup A_2 \cup A_3 \cup A_4$ if and only if $a \in A_j$ for some $j = 1, 2, 3$, or 4. In general, the intersection of a collection of sets consists of those elements that are in all the sets, while the union consists of those elements that are in at least one of them.

0.0 PROBLEMS

1. Let $A = \{0, 1, 2, 3, 4, 5\}$, $B = \{0, 3, 6, 9\}$, and $C = \{1, 9\}$. Find:
   (a) $A \cap B$
   (b) $A \cup C$
   (c) $B \cup (A \cap C)$
   (d) $(B \cup A) \cap C$
2. What are $A \cap \varnothing$ and $A \cup \varnothing$?
3. What can be said about the sets A and B if $A \cap B = A$?
4. State and prove a theorem like Problem 3 with $\cap$ replaced by $\cup$.
5. Prove that $\varnothing \subseteq B$ for any set B.
6. How many subsets does a set of 100 elements have? Prove your answer.
7. Using the usual notation for intervals on the line, find:
   (a) the intersection of all sets of the form $[0, x]$, $x > 0$
   (b) the intersection of all sets of the form $(0, x)$, $x > 0$
   (c) the union of all sets of the form $(-x, x)$, $x > 23$
   (d) the union of all sets of the form $[-x, x]$, $x > 23$

8. Let S be a set. If A and B are subsets of S, we define $A + B$ to be the set $\{x \mid x \in A \cup B \text{ and } x \notin A \cap B\}$.
   (a) What is $A + A$?
   (b) What is $A + \varnothing$?
   (c) Show that $(A + B) + C = A + (B + C)$ for any subsets A, B, and C of S.
   (d) Show that $A \cap (B + C) = (A \cap B) + (A \cap C)$ for any subsets A, B, and C of S.

CHAPTER I

INTEGERS

1.1 DIVISIBILITY

Mathematics starts with the *natural numbers* 1, 2, 3, 4, ... Early in the game, it was realized that life became much simpler if one also included 0 and the negative numbers -1, -2, -3, ...; all these numbers were known as "whole" numbers, to distinguish them from the little pieces—or fractions—of numbers, and, in accordance with the scientific community's penchant for names derived from Latin, are now called *integers*. We shall denote the set of integers by Z, a tribute to the German word *Zahl*, meaning number.

There are two basic operations that one can perform in Z: addition and multiplication. We denote the sum of two integers, a and b, by $a + b$ and their product by ab or $a \cdot b$. These operations are familiar ones, derived from counting. Notice that subtraction is always possible in Z; that is, for any $a, b \in Z$, the equation $b = x + a$ has a unique solution. This property, which the natural numbers lack (why?), can be regarded as the *raison d'être* for Z in the first place. Division—that is, solution of the equation $ax = b$—is *not* always possible in Z (why not?).

The inability to divide gives the integers a richer structure under multiplication than they have under addition, because *every* integer can be written as 7 *plus* something, but only very special integers can be written as 7 *times* something (where the "something" is an integer). The following notation is used to describe the idea of "goes into evenly":

DEFINITION. If a and b are integers, we say that *a divides b*, and write $a \mid b$, if there exists an integer x such that $ax = b$.

Thus, 4 divides -12, or $4 \mid -12$, because $4 \cdot (-3) = -12$. However, 2 does not divide 7 because there is no integer x such that $2x = 7$. Notice that $1 \mid z$ for any integer z, since $1 \cdot z = z$. Also $-1 \mid z$ for any integer z (why?). Clearly (?), 1 and -1 are the only two integers with this property. On the other hand, every integer z divides 0 since $z \cdot 0 = 0$; while the only integer that 0 divides is 0, for if $0 \cdot x = b$ then $b = 0$. Notice that the x in the definition is not unique when $a = b = 0$.

The relation of divisibility in the integers is similar to the relation $\leq$ in the real numbers. The most outstanding property of the latter is *transitivity*: if $a \leq b$ and $b \leq c$, then $a \leq c$. This property is shared by divisibility.

THEOREM 1. Let a, b, and c be integers. If $a \mid b$ and $b \mid c$, then $a \mid c$.

Proof: If $a \mid b$, then there is an x in Z such that $b = ax$. Similarly, if $b \mid c$, then there is a y in Z such that $c = by$. Thus, $c = by = (ax)y = a(xy)$, and so $a \mid c$.

If a and b are real numbers such that $a \leq b$ and $b \leq a$, then $a = b$. This *antisymmetry* property is *not* shared by divisibility; for example,

$3 \mid -3$ and $-3 \mid 3$ but $3 \neq -3$. A little thought should convince you that if a and b are integers such that $a \mid b$ and $b \mid a$, then $a = b$ or $a = -b$. If $a \mid b$ and $b \mid a$, we say that a and b are *associates*.

Any integer can be written as a product of two integers, since we always have $a = a \cdot 1$ and $a = (-a) \cdot (-1)$. For some integers, like 7 and 11, these are the only ways that they can be written as products. Such numbers are worthy of a special name.

DEFINITION. An integer p, different from 1 and -1, is called a *prime*, if whenever $p = ab$, where a and b are integers, then either a or b is 1 or -1.

An example of a prime is 17, since the only ways of writing 17 as a product are $1 \cdot 17$, $(-1) \cdot (-17)$, $17 \cdot 1$, and $(-17) \cdot (-1)$. Similarly, -23 is a prime. On the other hand, 6 is not a prime because $6 = 2 \cdot 3$, and neither 2 nor 3 is 1 or -1. The first few positive primes are 2, 3, 5, 7, 11, 13, 17, 19, 23, 29, 31, 37, ... The negatives of these numbers are also primes (see Problem 8 in this section). Note that 1 and -1 are explicitly excluded from being primes.

The primes are the building blocks of the integers under multiplication—in any event, the primes cannot be built up from other integers by means of multiplication. Two questions arise: are there enough primes to build the other integers? and, can the same integer be built up in more than one way? The answer to the first question is "yes." (The second question will be answered in section 1.3.)

THEOREM 2. Every nonzero integer is either 1, -1, a prime, or a product of primes.

Proof: Suppose not. Let a be a nonzero integer of smallest absolute value, which is neither 1, -1, a prime, nor a product of primes. Since a is not 1, -1, or a prime, then $a = bc$ where neither b nor c is 1 or -1. Clearly then $|b|$ and $|c|$ are smaller than $|a|$. By the minimality of $|a|$, b and c must be primes or products of primes. Thus a is a product of primes. But a was chosen *not* to be a product of primes—a contradiction.

Two powerful techniques were employed in the proof of this theorem. The overall plan of the proof was *reductio ad absurdum*, or proof by contradiction. We wished to show that something was true, so we considered what would happen if it were false and wound up with a patent absurdity. Since a statement is either true or false, and we eliminated the possibility that it was false, it could only be true. A key step in the proof was the singling out of an integer of smallest absolute value satisfying specific conditions. We can do this because *the positive integers are well-ordered*—any nonempty set of positive integers contains a smallest element. To appreciate the significance of this fact, notice that the set of positive real numbers does not have a smallest element (if x is a positive real number, then $x/2$ is a smaller positive real number). In particular, if we have a set of nonzero integers, their absolute values form a set of positive integers, and so there is one of smallest absolute value.

The proof of Theorem 2 may seem unsatisfying to you. In fact, it is doubtful that many mathematicians think along these lines, although they very often write that way. The reason for doing this is that it provides a concise way of writing down reasonably airtight arguments. It also allows one to tread more surely in situations where the path is not so clear. What's going on in the proof is roughly this: take an integer a; if it's not a prime, you can chop it in two parts, so that $a = bc$; if b and c are primes, you've made it; otherwise, chop up b and c into smaller parts, and keep chopping until you get down to primes. You inevitably will end up with primes, because every time you chop a number the factors get smaller (in absolute value), until you can't chop anymore. The proof merely makes a judicious choice of what integer a to look at, and only one chop is then required.

If a and b are integers, we say that the integer c is a *common divisor* of a and b if $c \mid a$ and $c \mid b$. The integer 3 is a common divisor of 6 and -15. Among all the common divisors of a and b, the most informative one is the largest. The integers 18 and 30 have as common divisors the numbers $1, -1, 2, -2, 3, -3, 6$, and -6. We see that 6 is the *greatest common divisor* (GCD) of 18 and 30. If c is the greatest common divisor of a and b, we write $c = (a, b)$; thus $6 = (18, 30)$. Notice that every common divisor of 18 and 30 divides 6.

Two integers a and b are said to be *relatively prime* if $(a, b) = 1$. If a and b are relatively prime, then (since 1 is the *greatest* common divisor of a and b) the only integers that divide both a and b are 1 and -1. The integers 6 and 35 are relatively prime.

1.1 PROBLEMS

1. Determine all integers x such that:
   (a) $x \mid 60$ (b) $x \mid 35$ (c) $x \mid 17$ (d) $x \mid -256$

2. List the twenty-five positive primes less than 100.

3. Write the following integers as products of primes:
   (a) 1000 (b) 2316 (c) 1001 (d) 1111

4. Determine the greatest common divisors of the following pairs of integers:
   (a) 30 and 105 (b) 126 and 132 (c) 15 and -18
   (d) -6 and -8 (e) 1 and x

5. What is $(14, 14)$? $(-6, -6)$? (a, a)? Be sure that $(a, a) = (-6, -6)$ when $a = -6$.

6. What is $(7, 49)$? $(-6, 36)$? (a, a^2)? Be sure that $(a, a^2) = (-6, 36)$ when $a = -6$.

7. If p is a prime, what can be said about (p, x)?

8. Prove that p is a prime if and only if $-p$ is a prime.

9. Prove that p is a prime if and only if p has precisely four divisors.

10. Show that if $a \mid b$ and $a \mid c$, then $a \mid (b + c)$.

11. Show that if $b = (s, t)$, then $(s/b, t/b) = 1$.

12. Show that if $(a^2, b^2) = 1$, then $(a, b) = 1$.

13. Use the fact that the positive integers are well-ordered to prove the *induction principle*: If S is a set such that $1 \in S$ and, whenever $k \in S$ then $k + 1 \in S$, then S contains all the positive integers. (Consider the smallest positive integer that is not in S.)

14. Show that if a, b, s, and t are integers, and $sa + tb = 1$, then $(a, b) = 1$.

15. Let a and b be positive integers. Show that $a \mid b$ if and only if $(a, b) = a$.

1.2 IDEALS AND THE DIVISION ALGORITHM

In dealing with the integers, we shall be concerned primarily with their multiplicative structure: how they behave under multiplication. Consider the integer 6. Our chief concern with 6 is how it enters into the multiplicative scheme of things—that is, which integers it divides. The integers that 6 divides are $\ldots, -18, -12, -6, 0, 6, 12, 18, \ldots$, or, all integers of the form $6n$, where n is an integer. This set completely describes the properties of 6 as a divisor since 6 divides a number if and only if it is in this set. What is the set of numbers that -6 divides? You may readily ascertain that it is the same set; as far as dividing is concerned, there is no essential difference between the numbers 6 and -6.

Let I denote the set of integers divisible by 6. The set I has three important properties:

I_1. $0 \in I$.
I_2. If $x \in I$ and $y \in I$, then $x + y \in I$.
I_3. If $x \in I$ and $z \in Z$, then $zx \in I$.

These properties are simple consequences of what it means to be divisible by 6. We know that $0 \in I$ because $6 \cdot 0 = 0$, so $6 \mid 0$. If $x \in I$ and $y \in I$, then $x = 6x'$ and $y = 6y'$, so $x + y = 6x' + 6y' = 6(x' + y')$, and so $x + y \in I$. If $x \in I$ and $z \in Z$, then $x = 6x'$, so $zx = z6x' = 6zx'$, and so $zx \in I$.

There is nothing sacred about the number 6 (for us anyway). The same argument would work for 15 or -68 or any other integer. Thus, if I is the set of multiples of some integer m, then I satisfies I_1, I_2, and I_3.

We call *any* set of integers I that satisfies I_1, I_2, and I_3 an *ideal*. The name "ideal" arose in the same way that the name "imaginary" arose for describing the complex number i. In some important number systems, which we shall examine in Chapters 8 and 9, it was found that there were sets of elements which satisfied I_1, I_2, and I_3, and so "should" have been sets of multiples of some number but in fact were not. To remedy this, *ideal* numbers were invented to take care of these sets; eventually, the sets themselves came to be called ideals. An ideal which is in fact the set of multiples of some number is called a *principal ideal*. If I is the set of all multiples of the integer m, we say that *m generates I*.

Ideals arise naturally in many circumstances. At Lucky Bob's Casino, all transactions are effected by exchanging \$7 and \$11 chips. A \$7 or \$11 wager is easily settled. There is clearly no trouble if the amount in question is \$7, \$14, \$21, \$28, etc., or \$11, \$22, \$33, etc., or even \$25. A \$3 debt may be paid off by parting with two \$7 chips and receiving one \$11 chip in return. It is clearly a flexible setup. What amounts of money can change hands?

Any payment must be of the form $7s + 11t$ where s and t are integers (positive, negative, or zero). The integers s and t are, respectively, the number of \$7 and \$11 chips received; receiving -5 chips being interpreted as paying 5 chips. What integers can be written in the form $7s + 11t$? The set of all such integers forms an ideal. In fact: $0 = 7\cdot 0 + 11\cdot 0$; if $x = 7s + 11t$ and $y = 7s' + 11t'$, then $x + y = 7(s + s') + 11(t + t')$; and if $x = 7s + 11t$, then $zx = 7sz + 11tz$.

The most important algebraic fact about the integers is that every ideal is principal. To prove this, we first establish a result that was known to Euclid.

THE DIVISION ALGORITHM. If a and b are integers and $a \neq 0$, then there exist integers q and r such that

$$b = qa + r \qquad \text{and} \qquad 0 \leq r < |a|.$$

Proof: There certainly exist integers x such that $b - xa \geq 0$, for example, $x = -a|b|$. Among these integers, let q be one for which $b - qa$ is the smallest. Let $r = b - qa$. Then $0 \leq r < |a|$; for if not, then $0 \leq b - qa - |a| < b - qa$, and so either $0 \leq b - (q+1)a < b - qa$ or $0 \leq b - (q-1)a < b - qa$ (depending on whether a is positive or negative) contradicting the choice of q.

This is an example of a "bad" proof, in the sense that it definitely is not the way to convince anyone that the division algorithm is true. However, it is short and can be checked out step by step. To see what's going on, do Problem 9. Chances are that anyone writing down a proof of the division algorithm has that picture in mind regardless of what he writes.

The division algorithm simply formalizes what every schoolboy knows: that when you divide a positive integer b by a positive integer a, you get a quotient q and a remainder r that is smaller than a. Our description is a little fussier because we also allow either a or b (or both) to be negative. The important thing about r is that $|r| < |a|$; the fact that we can arrange to have $r \geq 0$ is nice but not essential in most applications (see the proof of Theorem 3 below). In a more general setting, like polynomials or Gaussian integers, where we have no notion of positiveness but do have the concept of absolute value, we can dispense with this refinement.

With this simple but important weapon in hand, we can prove:

THEOREM 3. Every ideal in the integers is principal.

Proof: Let I be an ideal. If $I = \{0\}$, then I consists of all the multiples of 0 and is thus principal. If $I \neq \{0\}$, choose a nonzero element $a \in I$ of smallest absolute value. Since I satisfies property I_3, every multiple of a is in I. We must show, conversely, that every element of I is a multiple of a. Let b be an element of I. Then $b = qa + r$ where $0 \leq r < |a|$. But then $r = b + (-q)a$. Now $b \in I$ by assumption; $(-q)a \in I$ by I_3; thus, $b + (-q)a \in I$ by I_2. But this says that $r \in I$. Since $0 \leq r < |a|$, if $r \neq 0$ we have a contradiction since a was chosen as a nonzero element of smallest absolute value. Hence $r = 0$, and so $b = qa$ is a multiple of a.

The proof of Theorem 3 is quite natural. We seek a number a that generates I. If every element of I were a multiple of a, then every nonzero element of I would have absolute value at least as big as $|a|$. So our best chance is to pick an element a whose absolute value is as small as possible. Now we want to show that every element b in I is divisible by this a, so we divide b by a and show that the remainder r is zero.

1.2 PROBLEMS

1. Prove that the set of multiples of the integer m is an ideal.
2. Show that the q and r in the division algorithm are unique—that is, show that if $b = q'a + r'$ with $0 \leq r' < |a|$, then $q = q'$ and $r = r'$.

3. Consider the set of all integers of the form $s40 + t6$, where s and t are integers. Show that this set is an ideal and find an integer that generates it.
4. Show that the integers a and b generate the same principal ideal if and only if $a = b$ or $a = -b$.
5. Show that the set of *all* integers is an ideal. Show that an integer m generates this ideal if and only if $m = 1$ or $m = -1$.
6. What number generates the ideal that arose at Lucky Bob's Casino? What is its significance to the patrons?
7. Show that if the ideals A and B are generated respectively by the integers a and b, then $A \subseteq B$ if and only if $b \mid a$.
8. If a and b are integers, then d is the *least common multiple* of a and b if d is the smallest positive integer such that $a \mid d$ and $b \mid d$. If A, B, and D are the ideals generated respectively by a, b, and d, show that $D = A \cap B$.
9. Prove the division algorithm "geometrically" by looking at the points $\ldots, -3a, -2a, -a, 0, a, 2a, 3a, \ldots$ on the line, noting that the distance between adjacent points is $|a|$ and observing that b must lie between two adjacent points.
10. Show that if a, b, q, and r are integers, and $b = qa + r$, then $(a, b) = (a, r)$.
11. Use Problem 10 (repeatedly) to find:
    (a) $(247, 323)$ (b) $(1073, 1247)$ (c) $(695789, 989497)$

1.3 UNIQUE FACTORIZATION

We know that every integer except 0, 1, and -1 is a prime or can be built up from the primes under multiplication. We turn now to the second question posed in section 1.1: can the same element be built up in more than one way? A superficial answer to this question is "yes." Indeed,

$$12 = 2 \cdot 2 \cdot 3 = 2 \cdot 3 \cdot 2 = 3 \cdot 2 \cdot 2 = (-2)(-2)3 = (-2)2(-3),$$

and so on. However, aside from the order and the replacement of some primes by their negatives, the decomposition is unique. No matter how we factor 36 into primes, we end up with two 2's and two 3's, and possibly some minus signs.

Unique factorization can be better appreciated by examining situations in which it does not hold. Suppose we were restricted to the (positive) even integers. We could still multiply, and the notion of prime would still make sense. What would the primes be? The number 2 would certainly be a prime since it cannot be written as the product of two even numbers; since $4 = 2 \cdot 2$, 4 is not a prime; more surprising is the fact that 6 *is* a prime, because 6 cannot be written as the product of two *even* numbers. Similarly, 10, 14, and 18 are primes whereas 8, 12, 16, and 20 are not. Now $36 = 6 \cdot 6 = 2 \cdot 18$ can be written as a product of primes in two very distinct ways.

The crucial problem in the preceding example is that the prime 6 divides the product $2 \cdot 18$ but does not divide either of the factors, 2 or 18 (18 cannot be written as 6 times an *even* number). This cannot happen in the integers. To prove this, we first establish a result of independent interest.

THEOREM 4. If d is the greatest common divisor of the integers x and y, then there exist integers s and t such that $sx + ty = d$.

Proof: Consider the set of *all* integers that can be written in the form $sx + ty$ for some integers s and t. Clearly (see 1.2, Problem 3), this set is an ideal and hence must consist of all multiples of some integer d which we can take to be positive (why?). Since $x = 1 \cdot x + 0 \cdot y$ and $y = 0 \cdot x + 1 \cdot y$, both x and y are in this ideal and so are multiples of d. Thus d is a common divisor of x and y. We know that $d = sx + ty$ for some integers s and t since d is in the set of all such numbers. If e is a common divisor of x and y, then e must divide $d = sx + ty$. Thus, since d is positive, e cannot be greater than d. Therefore, $d = sx + ty$ is the greatest common divisor of x and y.

Remark: We have shown, *en passant*, that every common divisor of x and y divides the greatest common divisor.

In particular, if $(x, y) = 1$, then there exist integers s and t such that $sx + ty = 1$. We use this fact to prove the fundamental property of primes in the integers.

THEOREM 5. If p is a prime integer and $p \mid ab$, then $p \mid a$ or $p \mid b$.

Proof: Suppose $p \mid ab$ but $p \nmid a$ (p does not divide a). Then $(p, a) = 1$ (see 1.1, Problem 7). Therefore, there exist integers s and t such that $sp + ta = 1$. Multiplying both sides of this equation by b, we obtain $b = spb + tab$. But, since p divides both terms of the right-hand side of this equation, this implies that $p \mid b$.

We see that if a prime divides the product of two integers, then it must divide one of them. The same thing holds for the product of any number of integers.

COROLLARY. If p is a prime integer and $p \mid a_1 a_2 \cdots a_n$, then $p \mid a_j$ for some j.

Proof: Since $p \mid a_1(a_2 \cdots a_n)$ then $p \mid a_1$ or $p \mid a_2 \cdots a_n$. If $p \mid a_1$ we are through; otherwise, $p \mid a_2 (a_3 \cdots a_n)$ and so $p \mid a_2$ or $p \mid a_3 \cdots a_n$. If $p \mid a_2$ we are through; otherwise, $p \mid a_3 (a_4 \cdots a_n)$. Continuing, we conclude that $p \mid a_1$ or $p \mid a_2$ or $\cdots$ or $p \mid a_n$.

To simplify the statement of the main theorem, we consider a prime to be a product of (one) primes.

THE FUNDAMENTAL THEOREM OF ARITHMETIC. If a is an integer greater than 1, then a is a product of positive primes. This product is unique (except for the order of the factors).

Proof: We know, by Theorem 2, that the first part of this theorem is true. Suppose the second part were false. Then there would exist positive integers that could be written as the product of positive primes in two different ways. Let a be the smallest such positive integer. Then $a = p_1 p_2 \cdots p_n = q_1 q_2 \cdots q_m$ where the p's and q's are positive primes and the two decompositions are different. Since p_1 is a prime that divides a, p_1 must divide one of the q's (since it divides their product). Reordering the q's, if necessary, we may assume that $p_1 \mid q_1$. But q_1 is also a positive prime—so p_1 must equal q_1. Cancelling p_1 and q_1 gives us $p_2 \cdots p_n = q_2 \cdots q_m$. But this is a positive integer smaller than a—so *these* two decompositions must be the same! Hence, since $p_1 = q_1$, the original decompositions are the same—a contradiction.

1.3 PROBLEMS

1. Let $a = p_1^{n_1} \cdots p_s^{n_s}$ and $b = p_1^{m_1} \cdots p_s^{m_s}$ where the p's are distinct positive primes and the n's and m's are nonnegative integers (with the usual convention that $p^0 = 1$). Show that $a \mid b$ if and only if $n_j \leq m_j$ for $j = 1, 2, \ldots, s$.
2. (Euclid) Prove that the number of primes is infinite (if $p_1, \ldots, p_n$ were a complete list of the positive primes, consider the number $p_1 p_2 \cdots p_n + 1$).
3. For the following pairs of integers x, y, find integers s and t such that $sx + ty = (x, y)$.
   (a) 12 and 15 (b) 35 and 37 (c) 16 and 36
   (d) 0 and 6 (e) -1 and 6 (f) 8 and 8
4. Let d be the greatest common divisor of the three integers a, b, c. Show that there exist integers s, t, u such that $sa + tb + uc = d$. Generalize to more than three integers.
5. (Irrationality of $\sqrt{2}$) Show that there are no nonzero integers a and b such that $a^2 = 2b^2$ (look at the prime decompositions of both sides).
6. Show that every positive fraction can be reduced uniquely to lowest terms—that is, if a and b are positive integers, then there exist unique positive integers s and t such that $a/b = s/t$ and $(s, t) = 1$. (To prove existence, choose the smallest a for which they don't exist—you cancel until you can't any more. Use the fundamental theorem of arithmetic to prove uniqueness.)
7. Show that $(x, y) = 1$ if and only if $(x^2, y^2) = 1$. More generally, show that $(x^2, y^2) = (x, y)^2$.
8. Show that $d = \pm (x, y)$ if and only if $d \mid x$ and $d \mid y$ and, if $c \mid x$ and $c \mid y$, then $c \mid d$. Thus, the latter condition determines

(x, y) except for its sign. It has the advantage of being phrased entirely in terms of divisibility (compare: (x, y) is the *greatest* common divisor of x and y) allowing it to be used in situations where *greatest* makes no sense (for example, which is greater, 1 or i?).

9. Show that if $(n, a) = 1$ and $n \mid ab$, then $n \mid b$.

10. Show that if $(a, b) = 1$, $a \mid n$, and $b \mid n$, then $ab \mid n$. Can the condition "$(a, b) = 1$" be omitted here? Why or why not?

11. Let S be a set of integers that is closed under multiplication (that is, if $a, b \in S$ then $ab \in S$). We can talk about factorization and primes in S just like we did for the even integers. Find the first few primes, and discuss unique factorization, when S is:
    (a) the odd positive integers: 1, 3, 5, 7, 9, 11, . . .
    (b) the integers greater than 3: 4, 5, 6, 7, 8, 9, . . .
    (c) the positive integers of the form $3n + 1$: 1, 4, 7, 10, 13, 16, . . .

CHAPTER II

RESIDUE CLASS ARITHMETIC

2.1 CONGRUENCE

In the theory of numbers, great emphasis is placed on the concept of divisibility, and results are frequently couched in these terms. For example, Pierre de Fermat showed, in the 17th century, that if p is a prime and a is any integer not divisible by p, then $a^{p-1} - 1$ is divisible by p. Many other theorems were stated and proved in the course of time that involved statements of the form "$a - b$ is divisible by n." Carl Friedrich Gauss, considered by some to be the greatest mathematician of all time, recognized the importance of this notion and formalized it in the idea of *congruence of numbers*. This seemingly trivial notational innovation is one of the finest conceptual tools we have.

If n is a positive integer and a and b are integers with the property that $a - b$ is divisible by n, we say that a and b are *congruent modulo n*, and write $a \equiv b \pmod{n}$. Some examples of congruences are $3 \equiv 7 \pmod{4}$, $-17 \equiv 28 \pmod{3}$, $30 \equiv 0 \pmod{5}$. The parenthetical expression "mod n" is to be thought of as modifying the "$\equiv$" and not the "b"; although "$a \equiv b \pmod{n}$" is commonly read "a is congruent to b, modulo n," it would be more accurately read as "a is congruent, modulo n, to b."

If a is an integer and n is a positive integer, then according to the division algorithm, there are integers q and r such that $a = qn + r$ and $0 \leq r < n$. The number r, which is unique (see 1.2, Problem 2), is sometimes called the *remainder* or *residue* when a is divided by n. This number determines how a behaves modulo n in the following sense:

THEOREM 6. Two integers a and b are congruent modulo n if and only if their remainders when divided by n are equal.

For example, $9 \equiv 33 \pmod{4}$ because $9 = 2 \cdot 4 + 1$ and $33 = 8 \cdot 4 + 1$; $8 \equiv -1 \pmod{3}$ because $8 = 2 \cdot 3 + 2$ and $-1 = (-1)3 + 2$; $17 \equiv 29 \pmod{6}$ because $17 = 2 \cdot 6 + 5$ and $29 = 4 \cdot 6 + 5$.

Proof: Suppose $a = qn + r$ and $b = q'n + r'$ where $0 \leq r < n$ and $0 \leq r' < n$. Then $a - b = (q - q')\,n + r - r'$ and so $n \mid (a - b)$ if and only if $n \mid (r - r')$. However, since $0 \leq r < n$ and $0 \leq r' < n$, then $n \mid (r - r')$ if and only if $r = r'$.

Either directly or from Theorem 6, we see that congruence modulo n is an *equivalence relation*; in other words, it satisfies the three properties

(1) (reflexive) $a \equiv a \pmod{n}$
(2) (symmetric) If $a \equiv b \pmod{n}$, then $b \equiv a \pmod{n}$
(3) (transitive) If $a \equiv b \pmod{n}$ and $b \equiv c \pmod{n}$, then $a \equiv c \pmod{n}$.

If a is any integer, we may associate with a the set of all integers that are congruent to a modulo n. By Theorem 6, this is just the set of all integers that have the same remainder as a when divided by n. This set is called the *residue class* of a, modulo n. The set of even integers is the residue class of -6, modulo 2; $\{\ldots -8, -5, -2, 1, 4, 7, 10, \ldots\}$ is the residue class of 1, modulo 3. Notice that there are n residue classes modulo n, and that each integer is in precisely one of them.

The word *congruence* in algebra is reserved for those equivalence relations that behave well with respect to some algebraic operation. Congruence modulo n respects the arithmetic operations on the integers.

THEOREM 7. Suppose $a \equiv a' \pmod{n}$ and $b \equiv b' \pmod{n}$. Then

(1) $a + b \equiv a' + b' \pmod{n}$

(2) $a - b \equiv a' - b' \pmod{n}$

(3) $ab \equiv a'b' \pmod{n}$.

Proof: To prove statement 1 above, we must show that $(a + b) - (a' + b')$ is divisible by n. But $(a + b) - (a' + b') = (a - a') + (b - b')$, and both terms of this sum are divisible by n. A similar argument proves statement 2. To show that $ab - a'b'$ is divisible by n, we add and subtract ab'. Thus, $ab - a'b' = ab - ab' + ab' - a'b' = a(b - b') + (a - a')b'$, and again both terms are divisible by n.

Theorem 7 enables us to do simple arithmetic with the residue classes themselves. If A and B are residue classes, we can define $A + B$ to be the residue class of $a + b$, where a is any member of A and b is any member of B; by Theorem 7, we get the same result no matter which numbers a and b we choose. Similarly, we may define AB. For example, consider the residue classes modulo 7. Let A be the residue class of 2 and B the residue class of 4. Then $A + B$ would be the residue class of 6, or, equally well since 9 is congruent to 2 and 18 is congruent to 4, the residue class of $27 = 9 + 18$. Similarly, AB would be the residue class of 8 or of 162.

2.1 PROBLEMS

1. Show that $a \equiv 0 \pmod{n}$ if and only if $n \mid a$.
2. How can you tell when two numbers are congruent modulo 1? modulo 2? modulo 10?

3. Restate Fermat's theorem in terms of congruence.
4. Show that a number is divisible by 9 if and only if the sum of its digits (in base 10) is divisible by 9.
5. What can you say about a number that is congruent to 1 modulo n for all n? Prove your statement.
6. Show that if $a \equiv b \pmod{n}$ and $d \mid n$, then $a \equiv b \pmod{d}$.
7. Show that $m^2 \equiv 0$ or $1 \pmod{4}$ for any integer m.
8. Use Problem 7 to show that 10,003 is not the sum of two squares (of integers).
9. Show that $7^{100} + 1$ is not divisible by 3.

2.2 RESIDUE CLASS RINGS

If $n > 1$ is an integer we use Z_n to denote the set of residue classes modulo n. Once we have chosen such an n, we denote the residue class of an integer x modulo n by $\bar{x}$. This is a notational convenience to distinguish between integers and their corresponding residue classes. We will drop this notation when (hopefully) there is no danger of confusion. Thus, $x \equiv y \pmod{n}$ if and only if $\bar{x} = \bar{y}$.

The definitions of addition and multiplication in Z_n, given in the preceding section, amount to

$$\bar{x} + \bar{y} = \overline{x + y}$$
$$\bar{x} \cdot \bar{y} = \overline{xy}.$$

That is, to add the residue class of x to the residue class of y, pick an element in the residue class of x (and why not pick x?), add it to an element of the residue class of y (y will do), and take the residue class of the sum. Multiplication is defined similarly. Theorem 7 assures us that we get the same result no matter which elements we pick.

The residue class $\bar{0}$ behaves just like 0 does:

$$\bar{x} + \bar{0} = \overline{x + 0} = \bar{x}.$$

The residue class $\bar{1}$ behaves just like 1 does:

$$\bar{x} \cdot \bar{1} = \overline{x \cdot 1} = \bar{x}.$$

Nor does the similarity to ordinary arithmetic end there. We can also subtract in Z_n—in other words, we can always solve the equation

$$\bar{b} = \bar{x} + \bar{a};$$

in fact,

$$\bar{b} = \overline{b - a} + \bar{a}.$$

Almost all of the elementary manipulations in Z are also valid in Z_n.

THEOREM 8. Addition and multiplication in Z_n have the following properties:

A_1. $A + B = B + A$
A_2. $A + (B + C) = (A + B) + C$
A_3. $A + \bar{0} = A$
A_4. $A + X = \bar{0}$ for some X

M_1. $AB = BA$
M_2. $A(BC) = (AB)C$
M_3. $A \cdot \bar{1} = A$

D. $A(B + C) = AB + AC$

for all $A, B, C \in Z_n$.

Proof: These properties all follow immediately from the corresponding properties of the integers. As an example, we will prove A_2. Let a, b, and c be members of the residue classes A, B, and C respectively ($A = \bar{a}$, $B = \bar{b}$, and $C = \bar{c}$). Then $A + (B + C)$ $= \overline{a + (b + c)} = \overline{(a + b) + c} = (A + B) + C$.

Properties A_1 and M_1 are known as the *commutative laws*; properties A_2 and M_2 are the *associative laws*; we call $\bar{0}$ the *additive identity* because of A_3; similarly, $\bar{1}$ is the *multiplicative identity* because of M_3; A_4 says you can subtract (see 2.2, Problem 7); property D is known as the *distributive law*, the one law that relates addition and multiplication. Any mathematical system with operations $+$ and $\cdot$ that satisfy all these properties is called a *ring* or, more specifically, a commutative ring with identity. Roughly speaking, whenever you can add, subtract, and multiply in accordance with the usual rules of arithmetic, you're dealing with a ring.

Since $\bar{0}$ and $\bar{1}$ have no *a priori* meaning in the general situation, properties A_3 and M_3 are then interpreted as asserting the *existence* of such elements. In practice, it is usually obvious what $\bar{0}$ and $\bar{1}$ are, and they are written simply as 0 and 1. Problems 6 to 10 in this section are devoted to deriving additional properties of rings from these basic properties.

We have already used the ring properties of the integers. In the proof of Theorem 1, we relied on the associative law for multiplication (do you see where?). In verifying that the set of integers divisible by 6 was an ideal, we made use of the distributive law and the associative law for multiplication. These properties are so familiar that we use them without thinking.

Not all of the manipulations that we do with integers are valid in Z_n. The most striking difference is the inability to cancel nonzero factors in an equation. The following is always true in Z:

$$M_4. \qquad \text{If } a \neq 0 \text{ and } ab = ac, \text{ then } b = c.$$

This is not always true in Z_n. For example, $2 \cdot 1 \equiv 2 \cdot 4 \pmod 6$ and $2 \not\equiv 0 \pmod 6$, but $1 \not\equiv 4 \pmod 6$. A ring in which M_4 is satisfied is called an *integral domain.*

Let's take a closer look at these rings. Consider Z_2. There are two residue classes modulo 2—the even integers and the odd integers. We might call these two classes "even" and "odd." The rules for addition and multiplication can be stated as "even + even = even" (since the sum of an even number and an even number is even), "even + odd = odd" (since the sum of an even number and an odd number is odd), "even · odd = even," and so on. Since "even" and "odd" are cumbersome expressions to manipulate, let's abbreviate them as $\bar{0}$ and $\bar{1}$, since 0 and 1 are, respectively, the smallest nonnegative even and odd numbers (and the only possible remainders when dividing by 2). The rules of arithmetic in Z_2 then become

$$\begin{array}{ll} \bar{0} + \bar{0} = \bar{0} \qquad & \bar{0} \cdot \bar{0} = \bar{0} \\ \bar{0} + \bar{1} = \bar{1} & \bar{0} \cdot \bar{1} = \bar{0} \\ \bar{1} + \bar{0} = \bar{1} & \bar{1} \cdot \bar{0} = \bar{0} \\ \bar{1} + \bar{1} = \bar{0} & \bar{1} \cdot \bar{1} = \bar{1} \end{array}$$

These can be written in the form of tables as

+	$\bar{0}$	$\bar{1}$
$\bar{0}$	$\bar{0}$	$\bar{1}$
$\bar{1}$	$\bar{1}$	$\bar{0}$

·	$\bar{0}$	$\bar{1}$
$\bar{0}$	$\bar{0}$	$\bar{0}$
$\bar{1}$	$\bar{0}$	$\bar{1}$

In Z_3, there are three residue classes corresponding to remainders of 0, 1, and 2 when dividing by 3. As in the case of Z_2, we label the elements of Z_3 by $\bar{0}$, $\bar{1}$, and $\bar{2}$. The addition and multiplication tables are

$+$	$\bar{0}$	$\bar{1}$	$\bar{2}$
$\bar{0}$	$\bar{0}$	$\bar{1}$	$\bar{2}$
$\bar{1}$	$\bar{1}$	$\bar{2}$	$\bar{0}$
$\bar{2}$	$\bar{2}$	$\bar{0}$	$\bar{1}$

$\cdot$	$\bar{0}$	$\bar{1}$	$\bar{2}$
$\bar{0}$	$\bar{0}$	$\bar{0}$	$\bar{0}$
$\bar{1}$	$\bar{0}$	$\bar{1}$	$\bar{2}$
$\bar{2}$	$\bar{0}$	$\bar{2}$	$\bar{1}$

We may observe directly that M_4 is satisfied in Z_2 and Z_3: the only nonzero element of Z_2 is $\bar{1}$, which is always cancellable; that $\bar{2}$ is cancellable in Z_3 is evident upon looking at the $\bar{2}$ column in the multiplication table. These two integral domains satisfy an even stronger condition than M_4: they satisfy the multiplicative analog of A_4, the existence of multiplicative inverses. Because $\bar{0}$ can never have an inverse, we must exclude it from consideration.

M_4'. If $a \neq 0$, there is an element b such that $ab = 1$.

There are three things to check to show that both Z_2 and Z_3 satisfy M_4'; check them.

A ring that satisfies M_4' is called a *field*. Examples of fields are Z_2, Z_3, the real numbers, and the rational numbers.

THEOREM 9. Every field is an integral domain.

Proof: We must show that M_4 holds in any field. Suppose $a \neq 0$ and $ab = ac$ in a field. Then there is a t such that $ta = 1$. Hence $tab = tac$, so $b = c$.

If a and c are elements of a field F, and $a \neq 0$, then we can make sense out of the symbol c/a, just as we do for real numbers. By c/a we mean the solution of the equation $ax = c$. This equation has a solution in F, for if $ab = 1$, then $x = bc$ works. Moreover, there is only one solution, because if $ax = c = a(bc)$, then $x = bc$, since F is an integral domain. In this notation we would write $b = 1/a$ for the b in M_4'. A field is simply a ring in which you can always divide, except by 0.

We continue our study by examining Z_4. The addition and multiplication tables are easily computed (we omit the bars over the numbers):

+	0	1	2	3
0	0	1	2	3
1	1	2	3	0
2	2	3	0	1
3	3	0	1	2

·	0	1	2	3
0	0	0	0	0
1	0	1	2	3
2	0	2	0	2
3	0	3	2	1

Here is an example of a ring that is not an integral domain. By looking at the "2" column in the multiplication table, we see that $2 \cdot 1 = 2 \cdot 3$, so M_4 does not hold. Notice also that $2 \cdot 2 = 0$; the product of nonzero elements can be zero. An element a such that $ab = 0$ for some nonzero element b is called a *zero-divisor*. In the ring Z_4, there are two zero-divisors, 0 and 2. In an integral domain, the only zero-divisor is 0.

2.2 PROBLEMS

1. Show that A_1, M_2, and D are true in Z_n.
2. Write down the multiplication tables of Z_5 and Z_6. Find the zero-divisors in each ring.
3. Find all the zero-divisors in Z_{20}.
4. Verify that Z_{13} is a field. Find 1/2, 2/7, and 3/5 in this field.
5. Let R be the set of real numbers of the form $a + b\sqrt{2}$ where a and b are rational numbers. Check that both the sum and the product of two elements of R are still in R (ordinary addition and multiplication of real numbers). Show that with these operations, R is a field.
6. Show that if a, b, and c are elements of a commutative ring and $ab = ac = 1$, then $b = c$.
7. To subtract A from B means to solve the equation $Y + A = B$. Show that this is always possible in a ring (take $Y = B + X$ where $A + X = \bar{0}$, from A_4).
8. Show that the Y in Problem 7 is unique (if $Y + A = Z + A$, then $(Y + A) + X = (Z + A) + X$). We write $Y = B - A$ for this unique element Y. Conclude that both the $\bar{0}$ in A_3 and the X in A_4 are unique.
9. Show that $\bar{0} \cdot A = \bar{0}$ (note that $\bar{0} \cdot A = (\bar{0}+\bar{0})A = \bar{0} \cdot A + \bar{0} \cdot A$ and use Problem 8).

10. Derive the subtractive distributive law: $A(B-C) = AB - AC$ (show that $A(B-C) + AC = AB$).
11. Show that a ring is an integral domain if and only if 0 is the only zero-divisor.
12. Show that if a and b are elements of a field, and $b \neq 0$, then $a/b = a \cdot (1/b)$.
13. Let F be a field, $a, b, c, d \in F$, $b \neq 0$, and $d \neq 0$. Show that $bd \neq 0$, $(a/b)(c/d) = ac/bd$, and $a/b + c/d = (ad + bc)/bd$.

2.3 THE GROUP OF UNITS

Recall that two nonzero integers are said to be *relatively prime* if their greatest common divisor is 1—that is, if the only integers that divide both of them are 1 and -1. This notion is intimately connected with that of zero-divisors in Z_n.

THEOREM 10. If n and a are integers with $n > 1$, then a and n are relatively prime if and only if the residue class of a modulo n is not a zero-divisor in Z_n.

Proof: Suppose that a and n are relatively prime and $ab \equiv 0 \pmod{n}$. By Theorem 4, there are integers s and t such that $sa + tn = 1$. Multiplying both sides of this equation by b, we get $sab + tnb = b$, so $b \equiv 0 \pmod{n}$. Alternatively, use Problem 9 of section 1.3.

Conversely, suppose $1 < c = (a, n)$. Then

$$(n/c)a = n(a/c) \equiv 0 \pmod{n}$$

but $n/c \not\equiv 0 \pmod{n}$; thus, a is a zero-divisor modulo n.

COROLLARY. Z_p is an integral domain for all positive primes p.

Proof: If p is a prime, then p is certainly relatively prime to any integer that it does not divide. Hence the nonzero elements of Z_p are not zero-divisors. But if a is not a zero-divisor and $ab = ac$, then $a(b - c) = 0$, so $b - c = 0$, or $b = c$.

If R is a commutative ring and $a \in R$ is such that, for some b, $ab = 1$, then a is called a *unit*. It is easy to show (2.2, Problem 6) that at most one such b exists. This unique element (if there is one) is denoted by a^{-1} and is called the (multiplicative) inverse of a. Thus, units are those elements that have inverses. A unit can never be a zero-divisor, for if a is a unit and $ab = 0$, then $b = a^{-1}ab = a^{-1}0 = 0$. On the other hand, there are integers (2, for example) that are neither units nor zero-divisors; this can only happen when the ring is infinite, as the following theorem shows.

THEOREM 11. If R is a finite ring and $a \in R$, then a is either a unit or a zero-divisor.

Proof: Let $r_1, r_2, \ldots, r_n$ be a list of the elements of R. Consider the elements $ar_1, ar_2, \ldots, ar_n$. If a is not a zero-divisor, then these elements are all distinct, for if $ar_i = ar_j$, then $a(r_i - r_j) = 0$, so $r_i = r_j$. But R only has n elements, so $ar_1, ar_2, \ldots, ar_n$ must be a complete list of all of them. In particular, one of the ar_i must be 1, so a is a unit.

This proof employs a form of the famous "pigeonhole principle," which states that if n pigeons are put in n pigeonholes, and no pigeonhole contains *more than* one pigeon, then every pigeonhole contains *at least* one pigeon. Both the pigeons and the pigeonholes are the elements of R, and we put the pigeon r in the pigeonhole ar. The theorem is proved by observing that there must be a pigeon in pigeonhole 1. Notice how this argument depends heavily on the fact that R is finite; try it out on the infinite ring Z, using $a = 2$.

COROLLARY. Z_p is a field for every positive prime p.

Proof: In fact, Theorem 11 shows that any finite integral domain is a field: no element that is different from 0 is a zero-divisor, so every element that is different from 0 has a multiplicative inverse.

If R is any ring, the set of units of R has some nice properties which we list in Theorem 12.

THEOREM 12. Let R be a ring and U the set of units of R. Then

(0) If $a \in U$ and $b \in U$, then $ab \in U$.
(1) If $a, b \in U$, then $ab = ba$.
(2) If $a, b, c \in U$, then $a(bc) = (ab)c$.
(3) $1 \in U$ and $a \cdot 1 = a$ for any a in U.
(4) If $a \in U$, there is an element $b \in U$ such that $ab = 1$.

Proof: To prove property 0, we note that $ab(b^{-1}a^{-1}) = 1$. Parts 1, 2, and 3 follow immediately from the properties of multiplication in R. Part 4 follows directly from the definition of U.

Property 0 merely says that the result of multiplying two units is another unit. This property is often referred to as *closure*, and we say that the units are *closed* under multiplication. Properties 1, 2, 3, and 4 are exactly like properties A_1, A_2, A_3, and A_4 of a ring, except that the operation in question is multiplication and the identity element is called 1 instead of 0. These two disparate notions, the elements of a ring under addition and the units of a ring under multiplication, are brought under the same roof by the concept of an *abelian group*.

An abelian group is, first of all, a set in which we can combine elements by an operation to get another element. This operation may be called multiplication, or addition, or anything else that is natural or strikes one's fancy. The symbols most often used to indicate the operation are $+$ or $\cdot$ or simply juxtaposition. For neutrality, let's indicate the result of combining two elements a and b by $a * b$. So an abelian group consists of a set G and an operation $*$ so that for every two elements a and b in G, we get an element $a * b$. The properties we demand of the operation $*$ to justify the name abelian group are

(0) If $a \in G$ and $b \in G$, then $a * b \in G$.
(1) If $a, b \in G$, then $a * b = b * a$.
(2) If $a, b, c \in G$, then $a * (b * c) = (a * b) * c$.
(3) There is an element $e \in G$ such that $a * e = a$ for all $a \in G$.
(4) If $a \in G$, there is a $b \in G$ such that $a * b = e$.

The element e is called the *identity* of the group. If the operation $*$ is $+$, we often use the symbol 0 instead of e; if the operation $*$ is $\cdot$, we use the symbol 1. The element b alluded to in property 4 is called the *inverse* of a and is often written a^{-1}, although $-a$ is used if the operation is $+$. When there is no danger of confusion, the group operation is indicated by juxtaposition, and the identity by 1, to cut down on notation.

One useful fact about groups is that we can cancel factors in an equation. Explicitly, if $a * x = a * y$, then $x = y$. To see this, "multiply" both sides of the equation $a * x = a * y$ by the inverse of a. We then get $a^{-1} * (a * x) = a^{-1} * (a * y)$, so $(a^{-1} * a) * x = (a^{-1} * a) * y$, which gives $e * x = e * y$, so $x = y$.

If the operation is not indicated by $+$ we employ the usual exponential notation: $a^1 = a$, $a^2 = a * a$, $a^3 = a * a * a$, ... etc. Also, $a^0 = e$ and $a^{-n} = (a^{-1})^n$. You should check that the usual laws of exponents hold. If the operation is indicated by $+$, we use the notation $2a = a + a$, $3a = a + a + a$, ... etc. Also $0 \cdot a = 0$ and $(-n)a = n(-a)$.

Let's examine the groups of units in the various rings we have encountered. First, consider the ring of integers. There are only two units in this ring, 1 and -1. The inverse of 1 is 1 since $1 \cdot 1 = 1$; the inverse of -1 is -1 since $(-1)(-1) = 1$. This group has two elements. For reasons that will become apparent, it is often referred to as *the* two-element group.

The next ring we considered was Z_2. Here there is only one unit: 1. The group of units thus consists of just one element and the rather uninteresting operation $1 \cdot 1 = 1$. For obvious reasons, this is called the *trivial* group.

Z_3 is a little more interesting. Here the units are 1 and 2, and the operation is specified by the usual property of 1 and the equality $2 \cdot 2 = 1$. Notice the similarity to the group of units of the integers. In both cases, the groups consist of two elements, but the similarity does not end there. The element 2 in Z_3 behaves like the element -1 in the integers—that is, its square is 1; and of course the elements called 1 in both rings behave so much alike that we give them the same name. Both groups are instances of the same general form: two elements a and b, such that $a^2 = b^2 = a$ and $ab = ba = b$.

What about Z_4? Since 2 is a zero-divisor $(2 \cdot 2 = 0)$, the units are 1 and 3; 1 has its usual property and $3 \cdot 3 = 1$. The two-element group again!

Since 5 is a prime, Z_5 is a field (Corollary to Theorem 11), so the group of units consists of the nonzero elements: 1, 2, 3, 4. Since $2^2 = 4$, $2^3 = 2 \cdot 4 = 3$, and $2^4 = 2 \cdot 3 = 1$, all these elements can be written as powers of 2. The structure of the group is thus completely determined by the fact that its elements are 1, 2, 2^2, and 2^3, and that $2^4 = 1$; the rules for multiplication are immediate. If every element of a group G can be written as g^n, n an integer, (or ng, if the operation is $+$) for some fixed element $g \in G$, then G is said to be *cyclic*, and the fixed element g is called a *generator*. Thus, the units of Z_5 form a cyclic group under multiplication with 2 as a generator. Observe that 3 is also a generator but 4 is not.

We have already considered another four-element group: Z_4 under addition. The elements of Z_4 are 0, 1, 2, and 3, and since $0 = 1 + 1 + 1 + 1$, $1 = 1$, $2 = 1 + 1$, and $3 = 1 + 1 + 1$, we see that 1 is a generator. Furthermore, it is easy to check that 3 is also a generator but 2 and 0 are not. Again, this group is really no different from the group of units of Z_5: they both have the form 1, a, a^2, a^3 where $a^4 = 1$. The apparent difference is caused by the fact that in the group of units of Z_5, the operation is multiplication and the identity is 1, whereas in the additive group of Z_4, the operation is addition and the identity is 0. The one group is really the same as the other, relabeled. We shall return to this point in some detail in the next chapter.

2.3 PROBLEMS

1. Find all units in Z_{10} and give the inverse in each case. Do the same for Z_7.
2. Show that the equation $ax = b$ is solvable in Z_n whenever a is not a zero-divisor. Is this equation ever solvable when a *is* a zero-divisor? Can you determine precisely what conditions on a and b allow you to solve $ax = b$ in Z_n?
3. Solve $4x = 2$ in Z_{21}.
4. Solve $18x = 7$ in Z_{125}.
5. Is the group of units of Z_6 cyclic? Z_7? Z_8? Z_9?
6. Show that an abelian group has exactly one identity. (If e_1 and e_2 are both identities, what is $e_1 * e_2$?)
7. Show that an element a in an abelian group has exactly one inverse. (If b and c are both inverses of a, what is $b * a * c$?)
8. Suppose G is a two-element group (multiplicative). Show that G consists of two distinct elements a and b, such that $a^2 = b^2 = a$ and $ab = ba = b$.

9. Show that if a and b are elements of a commutative ring, and ab is a unit, then a and b are units.

10. Is it true that if a and b are units in a ring, then $a+b$ is a unit? Check it on the rings that you know.

11. Show that (under addition) Z is a cyclic group, that Z_n is a cyclic group for every n, and that the rational numbers do not form a cyclic group.

12. If G is an abelian group and $a, b, \in G$, show that $a^n a^m = a^{n+m}$, $(ab)^n = a^n b^n$, and $(a^n)^m = a^{nm}$ for all integers n and m. Rewrite these identities in additive notation.

13. *Wilson's theorem.* Let p be a positive prime. Show that $1 \cdot 2 \cdot \cdots \cdot (p-1) \equiv -1 \pmod{p}$. (Pair each element in the product with its inverse; observe that, in a field, only 1 and -1 are their own inverses.) Show that the conclusion is false if $p > 1$ is not a prime.

14. Show that if $a \in Z_n$, then a is a generator of (the additive group) Z_n if and only if a is a unit in (the ring) Z_n.

15. Show that Z_n is not an integral domain if n is not a prime.

2.4 THE THEOREMS OF FERMAT, LAGRANGE, AND EULER

We return now to the theorem of Fermat mentioned in the first section: if $a \not\equiv 0 \pmod{p}$, then $a^{p-1} \equiv 1 \pmod{p}$, for p a positive prime. First, let's examine a few examples. In Z_{11}, consider the powers of 2: 2, 4, 8, 5, 10, 9, 7, 3, 6, 1. By direct computation, we see that $2^{10} \equiv 1 \pmod{11}$. Notice that we have also shown that the group of units of Z_{11} is cyclic and that 2 is a generator for it. Now consider the powers of 3: 3, 9, 5, 4, 1. We see that $3^5 \equiv 1 \pmod{11}$ and that 3 is not a generator for the group of units of Z_{11}. However, Fermat's theorem still holds here, for $3^{10} = (3^5)^2 \equiv 1^2 \equiv 1 \pmod{11}$. The elements 3, 9, 5, 4, 1 form a group under the operation of multiplication in Z_{11}. This is easily verified because they are exactly the powers of 3 in Z_{11}. Such sets play an important role in the general theory and warrant a name.

DEFINITION. If G is a group with operation $*$ and identity e, and H is a set of elements of G such that

(1) $e \in H$.
(2) If $h_1, h_2 \in H$, then $h_1 * h_2 \in H$.
(3) If $h \in H$, then $h^{-1} \in H$.

then H is said to be a *subgroup* of G.

Notice that for a subset H of a group G to be a subgroup of G, H must be a group under the operation $*$ of G, and conversely. Another example of a subgroup is the set of even integers in the additive group of the integers.

Fermat's theorem concerns when a power of an element is equal to 1 in a group. If a is any element of a group, the *order* of a is defined to be the least positive integer n such that $a^n = 1$. If no such positive integer exists, we say that a has infinite order. The order of an element tells us exactly which powers are equal to 1.

THEOREM 13. If a is an element of order n, then $a^t = 1$ if and only if $n \mid t$.

Proof: If $n \mid t$, then $t = kn$, so $a^t = a^{kn} = (a^n)^k = 1^k = 1$. On the other hand, if $a^t = 1$, then we can write $t = qn + r$, where $0 \leq r < n$, so $1 = a^t = a^{qn+r} = a^{qn}a^r = a^r$, and so r must be 0 by the minimality of n. Thus $t = qn$, so $n \mid t$.

In light of Theorem 13, we may rephrase Fermat's theorem as follows: if p is a prime and a is an element of the group of units of Z_p, then the order of a divides $p - 1$. We can further rephrase this in terms of subgroups. If a is any element of a group G, the *subgroup generated by a* is defined to be the set of powers of a (positive, negative, and zero). You should verify that this is indeed a subgroup. An example is the subgroup $\{3, 9, 5, 4, 1\}$ of the group of units of Z_{11}, which is generated by 3.

THEOREM 14. The order of a group element a is equal to the number of elements in the subgroup generated by a.

Proof: If the order of a is infinite, then the elements $a, a^2, a^3, \ldots$ are all distinct; if they were not, then for some $i<j$, $a^i = a^j$. But multiplying by a^{-i} shows that $1 = a^{j-i}$, contradicting the fact that a is of infinite order. If n is the order of a, then the elements $a, a^2, a^3, \ldots a^{n-1}, 1$ are all distinct by the same sort of argument. Moreover, all powers of a are present since $a^n = 1$. (Note that $a^{-1} = a^{n-1}$.) Thus, this is the subgroup generated by a, and it has precisely n elements.

Fermat's theorem can now be phrased as follows: if p is a prime and a is an element of the group of units of Z_p, then the number of elements in the subgroup generated by a divides $p - 1$, the total number of elements in the group of units of Z_p. A more general theorem is due to Lagrange. By the *order* of a group we mean the number of elements in it.

THEOREM (Lagrange). If H is a subgroup of a finite group G, then the order of H divides the order of G.

Proof: Suppose H is of order t and the elements of H are $h_1, h_2, \ldots h_t$. We shall consider sets of elements of G of the form $\{gh_1, gh_2, \ldots gh_t\}$, where g is some element of G. Such a set will be denoted by gH and called a *coset* of H. Observe that every coset contains precisely t elements; if $gh_i = gh_j$, then, multiplying on the left by g^{-1}, we have $h_i = h_j$. The second important fact about cosets is that they are either identical or they are disjoint. Indeed, suppose $fH \cap gH \neq \varnothing$. Then $fh_i = gh_j$ for some i and j. Multiplying both sides on the right by h_i^{-1}, we get $f = gh$ where $h = h_j h_i^{-1}$ is in H. Thus, $fH = ghH$. But hH is just H since all its elements are in H, and it has the same number of elements as H (the pigeonhole principle). Hence $fH = ghH = gH$ (write these equalities out in terms of $h_1, h_2, \ldots h_t$). We have shown that if two cosets have an element in common, then they are equal. Finally, since $1 \in H$, every element of G is in some coset ($g \in gH$). Hence, we have partitioned G into a number of sets, each of which has t elements. By a primordial notion of divisibility, the number of elements in G is divisible by t.

COROLLARY 1. If G is a group of order n and $g \in G$, then the order of g divides n, and $g^n = 1$.

Proof: Let m be the order of g. By Theorem 14, m is the order of the subgroup generated by g. By Lagrange's theorem, $m \mid n$. If $n = mk$, then $g^n = g^{mk} = 1^k = 1$.

COROLLARY 2. (Fermat) If a is an integer that is not divisible by the prime p, then $a^{p-1} - 1$ is divisible by p.

Proof: Applying Corollary 1 to the group of units of Z_p, we find that $\bar{a}^{p-1} = \bar{1}$, so $a^{p-1} - 1$ is divisible by p.

Note that no use was made in the proof of Lagrange's theorem of property 1 of abelian groups. Thus Lagrange's theorem will hold in mathematical systems that just satisfy 2, 3, and 4. These systems are simply called *groups*.

From Theorems 10 and 11, we know that the group of units of Z_n can be represented by those positive integers that are less than n and relatively prime to n. If n is any integer greater than 1, we define $\phi(n)$ to be the number of positive integers less than and relatively prime to n. This function is called *Euler's ϕ-function*. Thus, the order of the group of units of Z_n is $\phi(n)$. The first few values of $\phi(n)$ are $\phi(2) = 1$, $\phi(3) = 2$, $\phi(4) = 2$, $\phi(5) = 4$, $\phi(6) = 2$, $\phi(7) = 6$. Clearly $\phi(p) = p - 1$ for any prime p (why?). We have the following generalization of Fermat's theorem.

THEOREM (Euler). If $n > 1$ is an integer and a is an integer that is relatively prime to n, then $a^{\phi(n)} \equiv 1 \pmod{n}$.

Proof: Translate the theorem to Z_n, and apply Lagrange's theorem.

2.4 PROBLEMS

1. Find all the subgroups of Z_{12} (under addition).
2. Find all the subgroups of the group of units of Z_{12}.

3. Verify that $\{0, 3, 6\}$ is a subgroup of Z_9 (under addition). Break up Z_9 into cosets of $\{0, 3, 6\}$.
4. Verify that $\{1, 5\}$ is a subgroup of the group of units of Z_{24}. Break up the group of units of Z_{24} into cosets of $\{1, 5\}$.
5. Check Euler's theorem for $n = 25$.
6. Find $\phi(n)$ for $n = 8, 9, 10, 11, 12, 13, 14, 15$.
7. Find the orders of all the elements in the group of units of Z_{15}.
8. Show that if G is a group and $a \in G$, then the subgroup generated by a is the smallest subgroup of G containing a (that is, it is a subgroup of G containing a, and it is a subset of any subgroup of G containing a).
9. Show that a group G is cyclic if and only if G is the subgroup of G generated by some element $g \in G$.
10. Show that every group of prime order is cyclic.
11. Show that a group of (finite) order n is cyclic if and only if it has an element of order n.
12. Let $n > 1$ be an integer and I the principal ideal generated by n. Show that I is a subgroup of Z (under addition). What are the cosets of I?
13. Show that the intersection of any collection of subgroups of a group is a subgroup.

CHAPTER III

DECOMPOSITIONS

3.1 ISOMORPHISM

We proceed to a more penetrating analysis of the structure of the group of units of Z_n. We shall denote this group by U_n.

Consider U_3. It consists of the two elements 1 and 2 and the operation $\cdot$ with the multiplication table

$\cdot$	1	2
1	1	2
2	2	1

Now consider U_4. It consists of the two elements 1 and 3 and the operation $\cdot$ with the multiplication table

$\cdot$	1	3
1	1	3
3	3	1

Notice the similarity between these two tables. If we replace 2 by 3 in the first table, we get the second. Thus, we can view the second group as the same as the first group except that the elements have been renamed; the second group is the first group with 2 renamed as 3.

Similarly, consider U_8 and U_{12}:

U_8

$\cdot$	1	3	5	7
1	1	3	5	7
3	3	1	7	5
5	5	7	1	3
7	7	5	3	1

U_{12}

$\cdot$	1	5	7	11
1	1	5	7	11
5	5	1	11	7
7	7	11	1	5
11	11	7	5	1

Again, replacing 3 by 5, 5 by 7, and 7 by 11 in the first table yields the second. Contrast this to U_5:

$\cdot$	1	2	3	4
1	1	2	3	4
2	2	4	1	3
3	3	1	4	2
4	4	3	2	1

No relabeling will yield this table from either of the others. In fact, we observe that the diagonals, from upper left to lower right, in the first two tables consist of all the same element while in the third, two different elements appear. These groups are distinct not merely because the names of the elements are different, but because the intrinsic structures differ.

As a last example, consider the (additive) group Z_4.

+	0	1	2	3
0	0	1	2	3
1	1	2	3	0
2	2	3	0	1
3	3	0	1	2

Does this group have the same structure as one of the three four-element groups we have just examined? To see that it does, we replace 0 by 1, 1 by 2, 2 by 4, and leave 3 alone. The resulting table is

·	1	2	4	3
1	1	2	4	3
2	2	4	3	1
4	4	3	1	2
3	3	1	2	4

But this is the same as the multiplication table of U_5; we have merely written the elements 1, 2, 3, and 4 in a different order.

To formalize the notion of two groups having the same structure, we introduce the concept of an *isomorphism*, from the Greek for "same form."

DEFINITION. If G and H are two groups with operations $*_G$ and $*_H$, respectively, then G and H are *isomorphic* if we can make correspond to every element g in G an element $f(g)$ in H such that

(1) If $g_1 \neq g_2$, then $f(g_1) \neq f(g_2)$.
(2) Every element h in H equals $f(g)$ for some g in G.
(3) $f(g_1 *_G g_2) = f(g_1) *_H f(g_2)$.

The function f is said to be an *isomorphism* from G to H.

The first property is simply that the isomorphism f must assign distinct elements of H to distinct elements of G. This property is often expressed by saying that f is *one-to-one* (1-1). The second property says that every element of H is assigned to *some* element of G (and by property 1, this element is unique). This property is expressed by saying that f is *onto*. Properties 1 and 2 imply that G and H have the same number of elements (why?). It is property 3 that says that the multiplication tables of the two groups are the same. If we look at a typical entry in the tables of G

$*_G$	g_1
g_2	$g_1 *_G g_2$

$*_H$	$f(g_1)$
$f(g_2)$	$f(g_1) *_H f(g_2)$

and H, we see that property 3 is exactly what we need in order for the correspondence f to convert the table of G into the table of H. The subscript on the $*$, telling you which group you are operating in, is customarily omitted because it is ugly and usually quite unnecessary. If G is isomorphic to H, we write $G \cong H$.

For the groups U_8 and U_{12}, we construct an isomorphism f as follows:

$$\begin{aligned} 1 &\xrightarrow{f} 1 \\ 3 &\xrightarrow{f} 5 \\ 5 &\xrightarrow{f} 7 \\ 7 &\xrightarrow{f} 11; \end{aligned}$$

that is, $f(1) = 1$, $f(3) = 5$, $f(5) = 7$, and $f(7) = 11$. The first two requirements for an isomorphism are clearly met. To check the third, we must test all 16 possible pairs (g_1, g_2). We will check two here and leave the rest for you.

$$f(3 \cdot 5) = f(7) = 11 = 5 \cdot 7 = f(3) \cdot f(5)$$

$$f(3 \cdot 3) = f(1) = \ \ 1 = 5 \cdot 5 = f(3) \cdot f(3).$$

Any group that is isomorphic to U_8 or U_{12} is known as *Klein's four-group*.

On the other hand, suppose we had an isomorphism f from U_8 to U_5. Notice that $a^2 = 1$ for *every* element a in U_8. So $f(a^2) = f(a) \cdot f(a) = f(1)$ for every element a in U_8. But by property 2, *every* element in U_5 can be written as $f(a)$ for some a in U_8. Hence, if b is in U_5, then $b^2 = f(1)$. However, for example, $3^2 = 4 \neq 1 = 4^2$ in U_5. Hence, no such isomorphism f can exist.

A well-known isomorphism exists between the group of positive real numbers under multiplication and the group of all real numbers under addition. To every positive real number x, we make correspond the real number $\log x$. Properties 1 and 2 of an isomorphism are not difficult to check. Property 3 is the useful property of log that $\log xy = \log x + \log y$.

When you encounter a new group, the first question to ask is whether or not it is isomorphic to some group with which you are already familiar. The simplest groups are the additive groups Z_n, which are cyclic groups.

THEOREM 15. If G is a cyclic group of order n, then G is isomorphic to Z_n.

Proof: Let g be a generator of G. Then the elements of G are exactly $1, g, g^2, g^3, \ldots g^{n-1}$ (we use the multiplicative notation for convenience). An isomorphism from G to Z_n is then given by

$$\begin{aligned} 1 &\longrightarrow \bar{0} \\ g &\longrightarrow \bar{1} \\ g^2 &\longrightarrow \bar{2} \\ g^3 &\longrightarrow \bar{3} \\ &\;\;\vdots \\ g^{n-1} &\longrightarrow \overline{n-1}. \end{aligned}$$

That this is an isomorphism follows immediately from the fact that g is of order n. Note the similarity to log, which takes 10^x to x.

3.1 PROBLEMS

1. Let G and H be groups (multiplicative), and let e_G and e_H be their respective identity elements. Suppose f is an isomorphism from G to H. Show that:
   (a) $f(e_G) = e_H$
   (b) $f(a^{-1}) = (f(a))^{-1}$ for every $a \in G$
   (c) the order of a equals the order of $f(a)$ for every $a \in G$
   (d) G is cyclic if and only if H is cyclic.
2. Show that isomorphism is an equivalence relation.
3. (a) Show that any two two-element groups are isomorphic (see 2.3, Problem 8).
   (b) Show that any two three-element groups are isomorphic.
4. Show that if p is a prime and G has order p, then G is isomorphic to Z_p.
5. Show that every four-element group is isomorphic to Z_4 or to U_8.
6. Determine which of the groups U_n, $n = 2, 3, \ldots 15$, are isomorphic.

3.2 DIRECT SUMS

Mathematicians are forever trying to reduce complicated objects to simpler ones, one example being writing numbers as products of primes. A more sophisticated example is the idea of introducing a coordinate system into the plane so that each point in the plane corresponds to a unique ordered pair of real numbers, or, equivalently, to a point on the x-axis and a point on the y-axis. Two-dimensional problems are thus reduced to one-dimensional problems.

A similar technique can be applied to groups. Consider $U_8 = \{1, 3, 5, 7\}$ and two of its subgroups, $\{1, 3\}$ and $\{1, 5\}$. Every element of U_8 can be written uniquely as a product of an element in $\{1, 3\}$ and an element in $\{1, 5\}$:

$$\begin{aligned} 1 &= 1 \cdot 1 \\ 3 &= 3 \cdot 1 \\ 5 &= 1 \cdot 5 \\ 7 &= 3 \cdot 5 \end{aligned}$$

We indicate this situation by writing $U_8 = \{1, 3\} \oplus \{1, 5\}$.

Another example is provided by $U_{15} = \{1, 2, 4, 7, 8, 11, 13, 14\}$. Consider the subgroups $\{1, 2, 4, 8\}$ and $\{1, 11\}$. Every element in U_{15} is uniquely the product of an element in $\{1, 2, 4, 8\}$ and an element in $\{1, 11\}$. In fact,

$$\begin{aligned} 1 &= 1 \cdot 1 \\ 2 &= 2 \cdot 1 \\ 4 &= 4 \cdot 1 \\ 8 &= 8 \cdot 1 \\ 11 &= 1 \cdot 11 \\ 7 &= 2 \cdot 11 \\ 14 &= 4 \cdot 11 \\ 13 &= 8 \cdot 11. \end{aligned}$$

We therefore write $U_{15} = \{1, 2, 4, 8\} \oplus \{1, 11\}$.

DEFINITION. If A and B are subgroups of a group G such that for every element g in G there are unique elements a in A and b in B so that $g = ab$, then we write $G = A \oplus B$ and say that G is the *direct sum* of A and B.

The usefulness of the notion of a direct sum rests partly on the fact that if we know what A and B are, then we know what $A \oplus B$ is. This vague statement is made precise by the following theorem.

THEOREM 16. If $G = A \oplus B$, $G' = A' \oplus B'$ and $A \cong A'$, $B \cong B'$, then $G \cong G'$.

Proof: Suppose f is an isomorphism from A to A' and h is an isomorphism from B to B'. We construct an isomorphism k from G to G' by defining

$$k(g) = f(a) \cdot h(b)$$

where $a \in A$, $b \in B$, and $g = ab$. First, observe that this definition makes sense; if $g \in G$, then $g = ab$ for *unique* elements $a \in A$ and $b \in B$. Let's carefully check the three properties of an isomorphism.

(1) If $g_1 \neq g_2$, then, writing $g_1 = a_1 b_1$ and $g_2 = a_2 b_2$ where $a_1, a_2 \in A$ and $b_1, b_2 \in B$, we must have $a_1 \neq a_2$ or $b_1 \neq b_2$ (otherwise $g_1 = g_2$). Suppose $a_1 \neq a_2$ (if $b_1 \neq b_2$, the argument is the same). Then $f(a_1) \neq f(a_2)$ since f is an isomorphism from A to A'. But then $k(g_1) = f(a_1) \cdot h(b_1) \neq f(a_2) \cdot h(b_2) = k(g_2)$ because $f(a_1)$, $f(a_2) \in A'$ and $h(b_1)$, $h(b_2) \in B'$ and every element of G' is *uniquely* the product of an element of A' and an element of B'.

(2) If $g' \in G'$, then $g' = a'b'$ where $a' \in A'$ and $b' \in B'$. Since f and h are isomorphisms, there exist elements $a \in A$ and $b \in B$ such that $f(a) = a'$ and $h(b) = b'$. But then if $g = ab$, we have $k(g) = f(a) \cdot h(b) = a'b' = g'$.

(3) Suppose $g_1 = a_1 b_1$, $g_2 = a_2 b_2$, where a_1, $a_2 \in A$ and b_1, $b_2 \in B$. Then $k(g_1 g_2) = k(a_1 b_1 a_2 b_2) = k(a_1 a_2 b_1 b_2) = f(a_1 a_2) \cdot h(b_1 b_2)$ $= f(a_1) f(a_2) h(b_1) h(b_2) = f(a_1) h(b_1) f(a_2) h(b_2) = k(g_1) \cdot k(g_2)$.

The proof of Theorem 16, with all its letters and subscripts, may obscure the essential simplicity of the idea behind the theorem. The point is that if $G = A \oplus B$, then an element $g \in G$ may be described by the pair of "coordinates" (a, b), where $g = ab$. To multiply two elements of G, you simply multiply the respective coordinates. Hence, the multiplication in G is completely determined by the multiplications in A and B; so if A' is a copy of A and B' is a copy of B, then $A' \oplus B' = G'$ will just be a copy of G.

We illustrate the application of Theorem 16 by considering U_{15} and U_{16}. We know that $U_{15} = \{1, 2, 4, 8\} \oplus \{1, 11\}$. On the other hand, $U_{16} = \{1, 3, 5, 7, 9, 11, 13, 15\}$, and if we consider the two subgroups $\{1, 3, 9, 11\}$ and $\{1, 7\}$, we see that $U_{16} = \{1, 3, 9, 11\} \oplus \{1, 7\}$. In fact,

$$\begin{aligned} 1 &= 1 \cdot 1 \\ 3 &= 3 \cdot 1 \\ 9 &= 9 \cdot 1 \\ 11 &= 11 \cdot 1 \\ 7 &= 1 \cdot 7 \\ 5 &= 3 \cdot 7 \\ 15 &= 9 \cdot 7 \\ 13 &= 11 \cdot 7. \end{aligned}$$

But $\{1, 2, 4, 8\}$ and $\{1, 3, 9, 11\}$ are both cyclic groups of order four and so are isomorphic. Similarly $\{1, 11\}$ and $\{1, 7\}$ are cyclic groups of order two and so are isomorphic. Hence $U_{15} \cong U_{16}$.

For actual computation, the following theorem is useful.

THEOREM 17. If A and B are subgroups of G such that $A \cap B = \{1\}$ and the order of A times the order of B is equal to the order of G, then $G = A \oplus B$.

Proof: Consider the set of products of the form ab, $a \in A$, $b \in B$. We first show that these products are all distinct. Indeed, if $a_1 b_1 = a_2 b_2$, where $a_1, a_2 \in A$ and $b_1, b_2 \in B$, then $a_2^{-1} a_1 = b_2 b_1^{-1}$. But $a_2^{-1} a_1 \in A$ and $b_2 b_1^{-1} \in B$. Hence, $a_2^{-1} a_1 = b_2 b_1^{-1} = 1$, so $a_1 = a_2$ and $b_1 = b_2$.

So the number of products ab is equal to the order of A times the order of B. But this is also equal to the order of G. Hence, every element of G is of the form ab for unique elements $a \in A$ and $b \in B$ (the pigeonhole principle again).

With this theorem at our command, once we found the subgroups $\{1, 3, 9, 11\}$ and $\{1, 7\}$ of U_{16}, we would know immediately that $U_{16} = \{1, 3, 9, 11\} \oplus \{1, 7\}$ since $\{1, 3, 9, 11\} \cap \{1, 7\} = \{1\}$ and U_{16} has $8 = 4 \cdot 2$ elements. Another example, using additive groups, is afforded by Z_6. Two subgroups of Z_6 are $\{0, 2, 4\}$ and $\{0, 3\}$. But

$\{0, 2, 4\} \cap \{0, 3\} = \{0\}$ (the *additive* identity), and Z_6 has $6 = 3 \cdot 2$ elements. Hence, $Z_6 = \{0, 2, 4\} \oplus \{0, 3\}$, a fact that is readily verified (do so!).

3.2 PROBLEMS

1. Of the groups $Z_2, Z_3, \ldots, Z_{20}$, which are direct sums of two subgroups?
2. Can you guess the general theorem underlying your discoveries in Problem 1? Can you prove it?
3. Of the groups $U_{17}, U_{18}, U_{19}, U_{20}$, and U_{21}, which are cyclic and which can be written as direct sums? Find generators for the cyclic ones. Write as direct sums those that can be.
4. Use Theorem 16 to show that U_{21} is isomorphic to U_{28}.
5. Use Theorem 16 to show that U_{35} is isomorphic to U_{45}.
6. Show that if $G = A \oplus B$, then $A \cap B = \{1\}$. (If $a = b \in A \cap B$, then $a \cdot 1 = 1 \cdot b$.)
7. Show that any group G is the direct sum of itself and the subgroup consisting of just the identity. If you went this route on Problem 1, restate that problem, and do it again.
8. Show that the intersection of any two nonzero subgroups (that is, subgroups different from $\{0\}$) of the additive group of rational numbers is nonzero. Conclude that the rational numbers cannot be written as a direct sum in a nontrivial way. (The trivial way is described in Problem 7.)
9. Show that U_n has a subgroup of order two if $n > 2$.

3.3 MORE DIRECT SUMS

Consider the group Z_{30}. Two subgroups are $\{0, 6, 12, 18, 24\}$ and $\{0, 5, 10, 15, 20, 25\}$. The first is cyclic of order five while the second is cyclic of order six. Since the intersection of these two subgroups is $\{0\}$ and the product of their orders is thirty, we have

$$Z_{30} = \{0, 6, 12, 18, 24\} \oplus \{0, 5, 10, 15, 20, 25\}$$

by Theorem 17. On the other hand, the subgroup {0, 5, 10, 15, 20, 25} has two subgroups, {0, 10, 20} and {0, 15}, whose intersection is {0} and whose orders multiply to give six, the order of the subgroup {0, 5, 10, 15, 20, 25}. Hence, by Theorem 17,

$$\{0, 5, 10, 15, 20, 25\} = \{0, 10, 20\} \oplus \{0, 15\}.$$

Now every element of Z_{30} can be written uniquely as the sum of an element in {0, 6, 12, 18, 24} and an element in {0, 5, 10, 15, 20, 25}. On the other hand, we have just seen that every element of {0, 5, 10, 15, 20, 25} is uniquely the sum of an element in {0, 10, 20} and an element in {0, 15}. Therefore, every element of Z_{30} is uniquely the sum of three elements: one from {0, 6, 12, 18, 24}, one from {0, 10, 20}, and one from {0, 15}. We indicate this situation by writing

$$Z_{30} = \{0, 6, 12, 18, 24\} \oplus \{0, 10, 20\} \oplus \{0, 15\}.$$

In this manner, we speak of the direct sum of three subgroups just as well as the direct sum of two. Another example is given by U_{33}. The elements of U_{33} are 1, 2, 4, 5, 7, 8, 10, 13, 14, 16, 17, 19, 20, 23, 25, 26, 28, 29, 31, and 32—twenty in all. A subgroup is provided by the powers of 2: {1, 2, 4, 8, 16, 32, 31, 29, 25, 17}. Another subgroup is {1, 10}. The intersection of these two subgroups is {1} and the product of their orders is twenty. Hence, Theorem 17 says

$$U_{33} = \{1, 2, 4, 8, 16, 32, 31, 29, 25, 17\} \oplus \{1, 10\}.$$

Now look at the subgroup {1, 2, 4, 8, 16, 32, 31, 29, 25, 17}. It contains as subgroups {1, 4, 16, 31, 25} (the powers of 4) and {1, 32}. These subgroups again satisfy the hypotheses of Theorem 17, so

$$\{1, 2, 4, 8, 16, 32, 31, 29, 25, 17\} = \{1, 4, 16, 31, 25\} \oplus \{1, 32\}.$$

Reasoning as we did with Z_{30}, we see that every element of U_{33} can be written uniquely as a product of three elements, one from {1, 4, 16, 31, 25}, one from {1, 32}, and one from {1, 10}. Hence we write,

$$U_{33} = \{1, 4, 16, 31, 25\} \oplus \{1, 32\} \oplus \{1, 10\}.$$

In general, we have:

DEFINITION. If G is a (multiplicative) group and $A_1, A_2, \ldots, A_n$ are subgroups of G such that every element of G can be written uniquely as a product $a_1a_2 \cdots a_n$ where $a_j \in A_j$ for j = 1, 2, ..., n, then we say that G is the *direct sum* of the subgroups $A_1, A_2, \ldots, A_n$ and write $G = A_1 \oplus A_2 \oplus \cdots \oplus A_n$.

If $G = A_1 \oplus A_2 \oplus \cdots \oplus A_n$ and $g = a_1a_2 \cdots a_n$ where $a_j \in A_j$, we say that a_j is the *component* of g in the subgroup A_j. The component of g in A_j is unique by the definition of direct sum. A useful relation between the order of g and the orders of the components of g is provided by:

THEOREM 18. If $G = A_1 \oplus A_2 \oplus \cdots \oplus A_n$ and $g \in G$, then the order of g is the least common multiple of the orders of the components of G.

Proof: Write $g = a_1a_2 \cdots a_n$ where $a_j \in A_j$, j = 1, 2, ..., n. Then, for any positive integer s, $g^s = a_1{}^s a_2{}^s \cdots a_n{}^s$. Since $a_j{}^s \in A_j$, the component of g^s in A_j is $a_j{}^s$. Since $1 = 1 \cdot 1 \cdots 1$, the component of 1 in A_j is 1. Hence, $g^s = 1$ if and only if $a_j{}^s = 1$ for j = 1, 2, ..., n. But this occurs precisely when s is a multiple of the order of a_j (Theorem 13). Thus, s must be a common multiple of the orders of all the components of g. The order of g must therefore be the least common multiple of the orders of its components.

The group U_{21} will serve to illustrate this theorem. First, we write U_{21} as a direct sum. One subgroup consists of the powers of 2, {1, 2, 4, 8, 16, 11}. Another is {1, 13}. By Theorem 17, we have

$$U_{21} = \{1, 2, 4, 8, 16, 11\} \oplus \{1, 13\}$$

since U_{21} has twelve elements (count them). But {1, 2, 4, 8, 16, 11} = {1, 4, 16} $\oplus$ {1, 8} by Theorem 17 again. Thus,

$$U_{21} = \{1, 4, 16\} \oplus \{1, 8\} \oplus \{1, 13\}.$$

Consider the element $11 \in U_{21}$. Now $11 = 4 \cdot 8 \cdot 1$, so its components are 4, 8, and 1. Looking at powers of these elements, we get

11	4	8	1
$11^2 = 16$	$4^2 = 16$	$8^2 = 1$	
$11^3 = 8$	$4^3 = 1$		
$11^4 = 4$			
$11^5 = 2$			
$11^6 = 1$			

The orders of the components of 11 are thus 3, 2, and 1, while the order of 11 is 6, the least common multiple of 3, 2, and 1.

On the other hand, consider the element $20 \in U_{21}$. We have $20 = 1 \cdot 8 \cdot 13$. Looking at powers of these elements, we have

20	1	8	13
$20^2 = 1$		$8^2 = 1$	$13^2 = 1$

The orders of the components of 20 are 1, 2, and 2, while the order of 20 is 2, the least common multiple of 1, 2, and 2.

3.3 PROBLEMS

1. Decompose the following groups into direct sums of as many subgroups as possible.

   (a) Z_{42} (b) Z_{60} (c) Z_{72} (d) U_{105}

2. Show that Z_{p^2} cannot be decomposed into a (nontrivial) direct sum of two subgroups if p is a prime.
3. Show that any abelian group of order 6 is cyclic.
4. Show that if $G = A \oplus B$ and A and B are cyclic groups whose orders are relatively prime, then G is a cyclic group.
5. Verify by direct computation that $U_{33} = \{1, 4, 16, 31, 25\} \oplus \{1, 32\} \oplus \{1, 10\}$. Verify that the order of 26 is the least common multiple of the orders of its components.
6. Let g and h be elements of $G = A_1 \oplus A_2 \oplus \ldots \oplus A_n$. Show that if g_i is the component of g in A_i, and h_i is the component of h in A_i, then $g_i h_i$ is the component of gh in A_i, and g_i^{-1} is the component of g^{-1} in A_i.

3.4 EXTERNAL DIRECT SUMS

We have considered the problem of decomposing a group into the direct sum of some of its subgroups. In this section, we look at the reverse problem: given groups $A_1, A_2, \ldots, A_n$, can we construct a group G with subgroups $H_1, H_2, \ldots, H_n$ so that $G = H_1 \oplus H_2 \oplus \cdots \oplus H_n$ and $A_1 \cong H_1$, $A_2 \cong H_2, \ldots, A_n \cong H_n$? The answer is "yes," and the construction is simplicity itself.

Let us reexamine the structure of a direct sum. Suppose $G = H_1 \oplus H_2 \oplus \cdots \oplus H_n$. This means that every element g of G can be written uniquely as $g = g_1 g_2 \cdots g_n$ where $g_1 \in H_1, g_2 \in H_2, \ldots, g_n \in H_n$. Thus, to specify an element $g \in G$, one can just specify its string of coordinates $(g_1, g_2, \ldots, g_n)$. If k is another element of G, then k can be written uniquely as $k_1 k_2 \cdots k_n$ where $k_j \in H_j$. If we multiply g times k, we get

$$gk = g_1 g_2 \cdots g_n \cdot k_1 k_2 \cdots k_n = (g_1 k_1)(g_2 k_2) \cdots (g_n k_n).$$

But $g_j k_j \in H_j$ (why?). Hence, the string of coordinates corresponding to gk is $(g_1 k_1, g_2 k_2, \ldots, g_n k_n)$. This is readily computed from the strings of coordinates of g and k. We can get a picture of G by considering strings $(h_1, h_2, \ldots, h_n)$ where $h_j \in H_j$, and by multiplying these strings coordinatewise:

$$(g_1, g_2, \ldots, g_n) \cdot (k_1, k_2, \ldots, k_n) = (g_1 k_1, g_2 k_2, \ldots, g_n k_n).$$

This motivates the following definition.

DEFINITION. If $A_1, A_2, \ldots, A_n$ are groups, then the (external) *direct sum* of $A_1, A_2, \ldots, A_n$ is the set of strings $(a_1, a_2, \ldots, a_n)$ with $a_1 \in A_1, a_2 \in A_2, \ldots, a_n \in A_n$, where multiplication is given by

$$(a_1, a_2, \ldots, a_n) \cdot (b_1, b_2, \ldots, b_n) = (a_1 b_1, a_2 b_2, \ldots, a_n b_n).$$

THEOREM 19. The (external) direct sum G of the groups A_1, $A_2, \ldots, A_n$ is a group. $G = H_1 \oplus H_2 \oplus \cdots \oplus H_n$ where H_i consists of strings of the form $(a_1, a_2, \ldots, a_n)$ such that $a_j = 1$ for $j \neq i$. Moreover, $H_i \cong A_i$, $i = 1, 2, \ldots, n$.

Proof: The proof that G is a group is automatic; we must check closure, the identity, the associative law, the commutative law, and inverses. Closure is immediate from the definition of multiplication of strings. The identity is the string $(1, 1, \ldots, 1)$ since

$$(1, 1, \ldots, 1) \cdot (a_1, a_2, \ldots, a_n) = (1 \cdot a_1, 1 \cdot a_2, \ldots, 1 \cdot a_n) = (a_1, a_2, \ldots, a_n).$$

The commutative and associative laws follow directly from the corresponding laws in $A_1, A_2, \ldots, A_n$ (check them). The inverse of $(a_1, a_2, \ldots, a_n)$ is $(a_1^{-1}, a_2^{-1}, \ldots, a_n^{-1})$ since

$$(a_1, a_2, \ldots, a_n) \cdot (a_1^{-1}, a_2^{-1}, \ldots, a_n^{-1}) = (1, 1, \ldots, 1),$$

the identity of G.

We may write the element $(a_1, a_2, \ldots, a_n)$ as

$$(a_1, 1, 1, \ldots, 1) \cdot (1, a_2, 1, \ldots, 1) \cdot (1, 1, a_3, \ldots, 1) \cdots (1, 1, 1, \ldots, a_n),$$

where $(1, 1, \ldots, a_i, \ldots, 1)$ is in H_i. It is clear that we can do this in only one way (is it?). The sets H_i are subgroups (why?). Thus, $G = H_1 \oplus H_2 \oplus \cdots \oplus H_n$. The correspondence $f(1, 1, \ldots, a_i, \ldots, 1) = a_i$ is an isomorphism from H_i to A_i (check it).

At the risk of a little confusion, we avoid encumbering ourselves with more notation by simply writing $G = A_1 \oplus A_2 \oplus \cdots \oplus A_n$ if G is the external direct sum of the groups $A_1, A_2, \ldots, A_n$.

Examples:

(1) $U_3 \oplus U_4$. This consists of pairs (a, b) where $a \in U_3$ and $b \in U_4$. A complete listing is $(1, 1)$, $(1, 3)$, $(2, 1)$, $(2, 3)$. Multiplication is coordinatewise; for example, $(1, 3) \cdot (2, 3) = (1 \cdot 2, 3 \cdot 3) = (2, 1)$ and $(2, 3) \cdot (2, 3) = (2 \cdot 2, 3 \cdot 3) = (1, 1)$.

(2) $Z_2 \oplus Z_2$. Here the operation is addition. The group consists of all pairs (a, b) where $a, b \in Z_2$. A complete list is $(0, 0)$, $(1, 0)$, $(0, 1)$, $(1, 1)$. Sample additions are $(1, 0) + (1, 1) = (1+1, 0+1) = (0, 1)$ and $(1, 1) + (1, 1) = (1+1, 1+1) = (0, 0)$. Since $Z_2 \cong U_3 \cong U_4$, this group is isomorphic to the group in the preceding example.

(3) $\mathscr{R} \oplus \mathscr{R}$ where $\mathscr{R}$ is the additive group of real numbers. The group $\mathscr{R} \oplus \mathscr{R}$ consists of all pairs of real numbers (a, b). The result of adding (a, b) to (c, d) is $(a+c, b+d)$. This group is isomorphic to $\mathscr{C}$, the additive group of complex numbers. The isomorphism is given by $f(a, b) = a+bi$. The proof that f is an isomorphism is left to the reader.

The notion of external direct sum gives us a convenient notation for describing abelian groups. For example, we know that $U_8 = \{1, 7\} \oplus \{1, 3\}$. But $\{1, 7\} \cong Z_2$ and $\{1, 3\} \cong Z_2$. Hence, $U_8 \cong Z_2 \oplus Z_2$, and this fact tells all about the behavior of U_8.

3.4 PROBLEMS

1. Show that $Z_6 \cong Z_2 \oplus Z_3$.
2. Show that $Z_{15} \cong Z_3 \oplus Z_5$.
3. Show that Z_4 is not isomorphic to $Z_2 \oplus Z_2$.
4. Find direct sums of Z_m's isomorphic to U_n for $n = 3, 4, \ldots, 25$. (For example: $U_7 \cong Z_6$, $U_8 \cong Z_2 \oplus Z_2$, $U_{15} \cong Z_2 \oplus Z_4$).
5. Show that:
   (a) $A \oplus B \cong B \oplus A$.
   (b) $(A \oplus B) \oplus C \cong A \oplus (B \oplus C) \cong A \oplus B \oplus C$.
   (c) $(A_1 \oplus \cdots \oplus A_n) \oplus B \cong A_1 \oplus \cdots \oplus A_n \oplus B$.

CHAPTER IV

THE STRUCTURE OF Z_n AND U_n

4.1 DECOMPOSING Z_n

So far, we have computed direct sums in a more or less haphazard manner; we chose reasonable-looking elements and considered the subgroups they generated, trying to get the right intersection and orders. In this section, we give a general method for decomposing the groups Z_n. In so doing, we shall pave the way to a better understanding of the groups U_n.

Let's look at Z_{30} again. We have observed, in section 3.3 of the last chapter, that

$$Z_{30} = \{0, 6, 12, 18, 24\} \oplus \{0, 5, 10, 15, 20, 25\},$$

so Z_{30} is the direct sum of a cyclic group of order five and a cyclic group of order six. Hence, by Theorem 16, $Z_{30} \cong Z_5 \oplus Z_6$. Thus, there is an isomorphism from Z_{30} to $Z_5 \oplus Z_6$.

On the other hand, there is a very natural function f from Z_{30} to $Z_5 \oplus Z_6$ staring us in the face. If we know the residue class of an integer x modulo 30, then we know its residue classes modulo 5 and modulo 6. Indeed, to know the residue class of x modulo 30 is to know x to within a multiple of 30, so we certainly know x up to a multiple of 5 or 6. For example, if $x \equiv 13 \pmod{30}$, then $x \equiv 13 \equiv 3 \pmod 5$ and $x \equiv 13 \equiv 1 \pmod 6$. The function f simply takes the residue class of x modulo 30 to the residue classes of x modulo 5 and modulo 6, as in $f(13) = (3, 1)$. A complete table for f is

$f(0) = (0, 0)$	$f(6) = (1, 0)$	$f(12) = (2, 0)$	$f(18) = (3, 0)$	$f(24) = (4, 0)$
$f(1) = (1, 1)$	$f(7) = (2, 1)$	$f(13) = (3, 1)$	$f(19) = (4, 1)$	$f(25) = (0, 1)$
$f(2) = (2, 2)$	$f(8) = (3, 2)$	$f(14) = (4, 2)$	$f(20) = (0, 2)$	$f(26) = (1, 2)$
$f(3) = (3, 3)$	$f(9) = (4, 3)$	$f(15) = (0, 3)$	$f(21) = (1, 3)$	$f(27) = (2, 3)$
$f(4) = (4, 4)$	$f(10) = (0, 4)$	$f(16) = (1, 4)$	$f(22) = (2, 4)$	$f(28) = (3, 4)$
$f(5) = (0, 5)$	$f(11) = (1, 5)$	$f(17) = (2, 5)$	$f(23) = (3, 5)$	$f(29) = (4, 5)$

That f is one-to-one and onto can be read directly from the table. The remaining property of an isomorphism is fairly apparent if we note that $f(\bar{x}) = (\bar{x}, \bar{x})$, where the first $\bar{x}$ is thought of as being in Z_{30}, the second in Z_5, and the third in Z_6. In fact, we have the following more general theorem.

THEOREM 20. Let $n = ab$ where a and b are positive integers such that $(a, b) = 1$, and denote by 1_m the residue class of 1 modulo m. Then the function

$$f(x \cdot 1_n) = (x \cdot 1_a, x \cdot 1_b) \qquad x \in Z$$

is an isomorphism between Z_n and $Z_a \oplus Z_b$.

Proof: First, we must verify that the definition of f even makes sense. The problem is that there are many choices of x that yield the same element $x \cdot 1_n$ in Z_n. However, if $x \cdot 1_n = y \cdot 1_n$, then $(x - y) \cdot 1_n = 0$. But the order of 1_n is n, so $n \mid (x - y)$, and hence $a \mid (x - y)$ and $b \mid (x - y)$, whereupon $(x - y) \cdot 1_a = 0$ and $(x - y) \cdot 1_b = 0$. Thus, $x \cdot 1_a = y \cdot 1_a$ and $x \cdot 1_b = y \cdot 1_b$, so $f(x \cdot 1_n) = f(y \cdot 1_n)$.

To show that f is $1-1$, suppose $f(x \cdot 1_n) = f(y \cdot 1_n)$—that is, $x \cdot 1_a = y \cdot 1_a$ and $x \cdot 1_b = y \cdot 1_b$. Then $a \mid (x-y)$ and $b \mid (x-y)$ so, since $(a, b) = 1$, $n \mid (x-y)$, and therefore $(x-y) \cdot 1_n = 0$, or $x \cdot 1_n = y \cdot 1_n$. Since Z_n and $Z_a \oplus Z_b$ have the same number of elements, f must be onto (if some element of $Z_a \oplus Z_b$ were not f of something, then some element would have to be f of two different things—but f is $1-1$—the pigeonhole principle again). Finally,

$$\begin{aligned} f(x \cdot 1_n + y \cdot 1_n) &= f((x+y) \cdot 1_n) = ((x+y) \cdot 1_a, (x+y) \cdot 1_b) \\ &= (x \cdot 1_a + y \cdot 1_a, x \cdot 1_b + y \cdot 1_b) = (x \cdot 1_a, x \cdot 1_b) \\ &\quad + (y \cdot 1_a, y \cdot 1_b) = f(x \cdot 1_n) + f(y \cdot 1_n). \end{aligned}$$

By repeated application of Theorem 20, or better by a proof exactly like the proof of Theorem 20, you can show:

COROLLARY. Let $n = a_1 a_2 \cdots a_s$ where the a_i are positive integers such that $(a_i, a_j) = 1$ for $i \neq j$ (we say that the a_i are *pairwise relatively prime*). Then the function

$$f(x \cdot 1_n) = (x \cdot 1_{a_1}, x \cdot 1_{a_2}, \ldots, x \cdot 1_{a_s}) \qquad x \in Z$$

is an isomorphism between Z_n and $Z_{a_1} \oplus Z_{a_2} \oplus \cdots \oplus Z_{a_s}$.

A simple consequence of this is the *Chinese remainder theorem*, which deals with the existence of solutions to sets of simultaneous congruences.

THEOREM (Chinese remainder theorem). If $a_1, a_2, \ldots, a_s$ are pairwise relatively prime positive integers, and $b_1, b_2, \ldots, b_s$ are any integers, then there exists an integer x such that

$$\begin{aligned} x &\equiv b_1 \pmod{a_1} \\ x &\equiv b_2 \pmod{a_2} \\ &\vdots \\ x &\equiv b_s \pmod{a_s}. \end{aligned}$$

The integer x is unique modulo n, where $n = a_1 a_2 \ldots a_s$.

Proof: Consider the element $(b_1 \cdot 1_{a_1}, b_2 \cdot 1_{a_2}, \ldots, b_s \cdot 1_{a_s})$ in $Z_{a_1} \oplus Z_{a_2} \oplus \cdots \oplus Z_{a_s}$. By the corollary to Theorem 20, there exists a unique $x \cdot 1_n \in Z_n$ such that $x \cdot 1_{a_1} = b_1 \cdot 1_{a_1}$, $x \cdot 1_{a_2} = b_2 \cdot 1_{a_2}, \ldots, x \cdot 1_{a_s} = b_s \cdot 1_{a_s}$, i.e., there exists an integer x, unique modulo n, such that $x \equiv b_1 \pmod{a_1}, \ldots, x \equiv b_s \pmod{a_s}$.

Theorem 20 allows us to write Z_n as a direct sum of groups of the form Z_{p^k}, where p is a prime. For example, $Z_{30} \cong Z_2 \oplus Z_3 \oplus Z_5$ and $Z_{140} \cong Z_4 \oplus Z_5 \oplus Z_7$. Beware of trying to decompose further by claiming $Z_4 \cong Z_2 \oplus Z_2$ (see Problem 5). Although $2 \cdot 2 = 4$, $(2, 2) = 2 \neq 1$, and so Theorem 20 does not apply.

4.1 PROBLEMS

1. Show that if $(a, b) = 1$, $a \mid z$, and $b \mid z$, then $ab \mid z$. Where is this used in this section?
2. Show that $Z_7 \oplus Z_{15} \cong Z_{21} \oplus Z_5$.
3. Show that $Z_{1000} \oplus Z_{33} \cong Z_{88} \oplus Z_{375}$.
4. Show that $Z_6 \oplus Z_{10} \cong Z_2 \oplus Z_{30}$.
5. Show that Z_{ab} is not isomorphic to $Z_a \oplus Z_b$ if $(a, b) \neq 1$ (consider the orders of elements in the two groups).
6. Find an integer x such that $x \equiv 6 \pmod 9$ and $x \equiv 13 \pmod{14}$. Find an integer y such that $y \equiv 0 \pmod 9$, $y \equiv 6 \pmod{10}$ and $y \equiv 6 \pmod{11}$.
7. Bob wears a green tie every Friday, a red shirt every third day, and an orange hat every fourth day. Use the Chinese remainder theorem to show that there are days when Bob wears all three.
8. Write a proof of the corollary to Theorem 20.

4.2 DECOMPOSING U_n

The function f constructed in Theorem 20 is more than just an isomorphism between the groups Z_n and $Z_a \oplus Z_b$. It is this "more" that enables us to use f to study the structure of U_n. Knowing that Z_n and

$Z_a \oplus Z_b$ behave the same way under addition tells us nothing about U_n, where the operation is multiplication. However, the isomorphism f also respects the multiplication in Z_n.

We have seen that $Z_a \oplus Z_b$ is a group under coordinatewise addition. Because Z_a and Z_b are *rings*, there is also a natural way to multiply in $Z_a \oplus Z_b$. The idea is exactly like external direct sums of groups.

DEFINITION. If $R_1, R_2, \ldots, R_n$ are rings, the *direct sum* $R = R_1 \oplus R_2 \oplus \cdots \oplus R_n$ is the set of all strings $(r_1, r_2, \ldots, r_n)$ where $r_1 \in R_1$, $r_2 \in R_2$, $\ldots$, $r_n \in R_n$. Addition and multiplication are defined in R by

$$(r_1, r_2, \ldots, r_n) + (s_1, s_2, \ldots, s_n) = (r_1 + s_1, r_2 + s_2, \ldots, r_n + s_n)$$

and

$$(r_1, r_2, \ldots, r_n) \cdot (s_1, s_2, \ldots, s_n) = (r_1 s_1, r_2 s_2, \ldots, r_n s_n).$$

That R is indeed a ring follows automatically from the ring properties of $R_1, R_2, \ldots, R_n$. The additive identity is $(0, 0, \ldots, 0)$ and the multiplicative identity is $(1, 1, \ldots, 1)$. The reader should check that at least a few of the ring properties hold in R.

Observe that if we forget about multiplication, R is just the direct sum of the additive groups $R_1, R_2, \ldots, R_n$ as defined previously.

Examples:

(1) $Z_3 \oplus Z_5$. The elements are pairs (a, b) where $a \in Z_3$ and $b \in Z_5$, just as in the direct sum of the *groups* Z_3 and Z_5. Now, however, we also have a multiplication. Samples are $(2, 3) \cdot (1, 4) = (2, 2)$; $(0, 1) \cdot (2, 3) = (0, 3)$; and $(2, 3) \cdot (2, 2) = (1, 1)$.

(2) $\mathscr{R} \oplus \mathscr{R}$ where $\mathscr{R}$ is the set of real numbers. Recall that as an additive group $\mathscr{R} \oplus \mathscr{R}$ is the same as the complex numbers $\mathscr{C}$. As a ring, $\mathscr{R} \oplus \mathscr{R}$ is quite distinct from $\mathscr{C}$. For example, $(3, 0) \cdot (0, 5) = (0, 0)$, so $\mathscr{R} \oplus \mathscr{R}$ contains nonzero zero-divisors while $\mathscr{C}$ does not. The multiplication in $\mathscr{R} \oplus \mathscr{R}$ is given by $(a, b) \cdot (c, d) = (ac, bd)$, while the multiplication in $\mathscr{C}$ is given by $(a, b) \cdot (c, d) = (ac - bd, ad + bc)$ (isn't that how you multiply complex numbers?).

If we have two rings, we consider them to have the same form, or to be isomorphic, if they behave the same way under both addition *and* multiplication. An isomorphism between rings is a one-to-one correspondence that respects *both* operations. Explicitly:

DEFINITION. A *ring isomorphism* between two rings R_1 and R_2 is a group isomorphism f between the additive groups R_1 and R_2, with the additional property that

$$f(x \cdot y) = f(x) \cdot f(y).$$

THEOREM 21. The group isomorphism constructed in Theorem 20 between Z_n and $Z_a \oplus Z_b$ is a ring isomorphism. More generally, the isomorphism between Z_n and $Z_{a_1} \oplus Z_{a_2} \oplus \cdots \oplus Z_{a_s}$ in the Corollary is a ring isomorphism.

Proof: An exercise to check your knowledge of the definitions.

An example may clarify the fuss. We know that $Z_{10} = \{0, 5\} \oplus \{0, 2, 4, 6, 8\}$ and $\{0, 5\} \cong Z_2$ under the isomorphism $0 \to 0$, $5 \to 1$, while $\{0, 2, 4, 6, 8\} \cong Z_5$ under the isomorphism $0 \to 0$, $2 \to 1$, $4 \to 2$, $6 \to 3$, $8 \to 4$; thus, $Z_{10} \cong Z_2 \oplus Z_5$ under the isomorphism

$$\begin{gathered}
0 = 0 + 0 \to (0, 0) \\
1 = 5 + 6 \to (1, 3) \\
2 = 0 + 2 \to (0, 1) \\
3 = 5 + 8 \to (1, 4) \\
4 = 0 + 4 \to (0, 2) \\
5 = 5 + 0 \to (1, 0) \\
6 = 0 + 6 \to (0, 3) \\
\text{etc.}
\end{gathered}$$

We may add in the left column by adding in the right. For example, to find $2 + 3$, we add $(0, 1) + (1, 4)$ and get $(1, 0)$, which corresponds to 5. This is the characteristic property of an isomorphism. However, this particular isomorphism does not respect multiplication. Indeed, $1 \cdot 2 = 2$ but $(1, 3) \cdot (0, 1) = (0, 3)$, which corresponds to 6. Theorem 21 says that there is an isomorphism that respects both operations. This enables us to prove:

THEOREM 22. If a and b are positive integers such that $(a, b) = 1$, then $U_{ab} \cong U_a \oplus U_b$. More generally, if $n = a_1 a_2 \cdots a_s$ where the a_i are pairwise relatively prime positive integers, then $U_n \cong U_{a_1} \oplus U_{a_2} \oplus \cdots \oplus U_{a_s}$.

Proof: We know that Z_{ab} is ring isomorphic to $Z_a \oplus Z_b$. Thus, U_{ab} is isomorphic to the group of units of $Z_a \oplus Z_b$ (see Problem 6 of this section). But when is an element (x, y) a unit in $Z_a \oplus Z_b$? There must exist an element (x', y') such that $(x, y) \cdot (x', y') = (1, 1)$, the multiplicative identity of $Z_a \oplus Z_b$. But this just means that $xx' = 1$ and $yy' = 1$—that x and y are in U_a and U_b, respectively. Hence, the group of units of $Z_a \oplus Z_b$ is precisely the (external) direct sum of U_a and U_b. The same argument is used to prove the more general assertion.

COROLLARY 1. Let ϕ be the Euler ϕ-function (section 2.4). If a and b are positive integers such that $(a, b) = 1$, then $\phi(ab) = \phi(a)\phi(b)$.

Proof: We know that $\phi(ab)$ is the order of U_{ab}. But $U_{ab} \cong U_a \oplus U_b$, so the order of U_{ab} is the product of the orders of U_a and U_b; that is, $\phi(a)\,\phi(b)$.

COROLLARY 2. If $n = p_1^{k_1} p_2^{k_2} \cdots p_s^{k_s}$ where the p_i are distinct positive primes and the k_i are positive integers, then $U_n \cong U_{p_1^{k_1}} \oplus U_{p_2^{k_2}} \oplus \cdots \oplus U_{p_s^{k_s}}$.

COROLLARY 3. With the same hypothesis as above, $\phi(n) = \phi(p_1^{k_1})\,\phi(p_2^{k_2}) \ldots \phi(p_s^{k_s})$.

Proof: Use the fact that $\phi(m)$ is the order of U_m, and Corollary 2. Alternatively, use repeated applications of Corollary 1.

The last corollary reduces the computation of the ϕ-function to computing $\phi(p^i)$ where p is a prime. This is relatively simple.

THEOREM 23. If p is a positive prime and i is a positive integer, then $\phi(p^i) = p^{i-1}(p-1)$.

Proof: We must count the number of positive integers less than p^i that are relatively prime to p^i—that is, that are not multiples of p. There are $p^i - 1$ positive integers less than p^i. Of these, the multiples of p are $p, 2p, 3p, \ldots, (p^{i-1}-1)p$, there being $p^{i-1}-1$ of them. Hence, there are $(p^i-1)-(p^{i-1}-1) = p^i - p^{i-1} = p^{i-1}(p-1)$ that are *not* multiples of p.

For example, $\phi(350) = \phi(2 \cdot 5^2 \cdot 7) = \phi(2)\phi(5^2)\phi(7) = 1 \cdot (5 \cdot 4) \cdot 6 = 120$. Also, $\phi(1000) = \phi(2^3 \cdot 5^3) = \phi(2^3)\phi(5^3) = (4 \cdot 1)(25 \cdot 4) = 400$.

4.2 PROBLEMS

1. Prove Theorem 21.
2. Prove Corollary 2 to Theorem 22.
3. Construct an isomorphism between U_{13} and U_{26}.
4. Find direct sums of Z_m's isomorphic to U_{30}, U_{40}, U_{45}, and U_{63}.
5. Compute $\phi(69)$, $\phi(132)$, $\phi(245)$, $\phi(3600)$.
6. Show that if two rings are isomorphic, then their groups of units are isomorphic.
7. Show that if $\phi(n) = n-1$, then n is prime.
8. Show that if $(a, b) \neq 1$, then $\phi(ab) \neq \phi(a)\phi(b)$.
9. Show that $\phi(n) = n\left(1 - \frac{1}{p_1}\right)\left(1 - \frac{1}{p_2}\right) \cdots \left(1 - \frac{1}{p_s}\right)$ where $n = p_1^{e_1} p_2^{e_2} \cdots p_s^{e_s}$ is the unique factorization of n into positive primes.

4.3 POLYNOMIALS OVER A FIELD

If n is any positive integer, the group U_n decomposes into a direct sum of groups of the form U_q where q is a power of a prime. Except for some funny business with the prime 2, we shall show that these groups are all cyclic, so the structure of U_n is completely known—that is, in the sense that questions about U_n may be reduced to questions about cyclic groups.

We start by showing that U_p is cyclic if p is a prime, a consequence of a more general fact about fields. In the course of the argument, we shall be concerned about the number of solutions to certain equations in a field. Now any equation can be put into the form of some expression equaling zero. The expressions that arise most naturally in algebraic questions are *polynomials*.

A polynomial over a field F is an expression of the form

$$a_n x^n + a_{n-1} x^{n-1} + \cdots + a_1 x + a_0$$

where n is a nonnegative integer, the a_i are elements of F, and x is an "indeterminate" or "variable." The polynomial is determined by these *coefficients* $a_0, a_1, \ldots, a_n$; two polynomials are equal if their coefficients are equal, with the understanding that powers of x that do not appear are treated as if they appeared with a coefficient of 0. If x is replaced by an element α of the field F, we get $a_n\alpha^n + a_{n-1}\alpha^{n-1} + \cdots + a_1\alpha + a_0$, which is an element of F. Thus, every polynomial determines a function from F to F. However, two polynomials may have the same value for every replacement of x by an element of F, yet be distinct. For example, if $F = Z_p$ then $\alpha^p = \alpha$ for every element $\alpha \in F$, but the polynomials x^p and x are not the same.

The *degree* of a polynomial is the highest power of x that appears with a nonzero coefficient; this coefficient is called the *leading coefficient*. If the leading coefficient is 1, we say that the polynomial is *monic*. Thus, the polynomial $x^3 + x + 1$ has degree 3 and is monic, while $2x^3 + 4x + 6$ has degree 3 and is not monic. If every coefficient is zero, we do not talk about degree. The polynomial $x - 3$ has degree 1, while the polynomial 5 has degree 0. Polynomials of degree 0 are called *constant* polynomials, or simply *constants*; they correspond to the nonzero elements of the field F.

The way we write polynomials tells us how to add and multiply them, using the usual arithmetic properties and the laws of exponents. It is readily seen that if $f(x)$ is a polynomial of degree n and $g(x)$ is a polynomial of degree m, then $f(x)g(x)$ is a polynomial of degree $n+m$. There is a division algorithm for polynomials, just like for the integers.

THEOREM 24. If F is a field and $a(x)$ and $b(x)$ are polynomials with coefficients in F, with $a(x) \neq 0$, then there exist polynomials $q(x)$ and $r(x)$ with coefficients in F such that

$$b(x) = q(x)a(x) + r(x)$$

where $r(x) = 0$, or the degree of $r(x)$ is less than the degree of $a(x)$.

Proof: Fix $a(x)$. We shall show that for any $b(x)$ we can find the required $q(x)$ and $r(x)$. Suppose not. Choose a $b(x)$ of smallest degree for which we cannot. The degree of $b(x)$ cannot be less than the degree of $a(x)$ for, were this so, $q(x) = 0$ and $r(x) = b(x)$ would work. Let $b(x) = b_n x^n + \cdots + b_1 x + b_0$ and $a(x) = a_m x^m + \cdots + a_1 x + a_0$ where $n \geq m$, $b_n \neq 0$ and $a_m \neq 0$. Then the polynomial

$$b(x) - \frac{b_n}{a_m} a(x)x^{n-m}$$

has degree less than n, the degree of $b(x)$. Hence, by the minimality of the degree of $b(x)$, we can find $q'(x)$ and $r(x)$ such that

$$b(x) - \frac{b_n}{a_m} a(x)x^{n-m} = q'(x)a(x) + r(x)$$

where $r(x) = 0$, or the degree of $r(x)$ is less than the degree of $a(x)$. Hence, if we let

$$q(x) = q'(x) + \frac{b_n}{a_m} x^{n-m}$$

we have $b(x) = q(x)a(x) + r(x)$, as required.

In practice, just like for the integers, we find $q(x)$ and $r(x)$ by "dividing" $b(x)$ by $a(x)$. For example, if $F = Z_7$ and $b(x) = 3x^4 - 2x^3 + x^2 + 5x - 4$, $a(x) = 3x^2 + x - 2$, we proceed by the format

$$\begin{array}{r|l}
 & x^2 - x + 6 \\
\hline
3x^2 + x - 2 & 3x^4 - 2x^3 + x^2 + 5x - 4 \\
 & 3x^4 + x^3 - 2x^2 \\
\cline{2-2}
 & - 3x^3 + 3x^2 + 5x \\
 & - 3x^3 - x^2 + 2x \\
\cline{2-2}
 & 4x^2 + 3x - 4 \\
 & 4x^2 + 6x - 5 \\
\cline{2-2}
 & 4x + 1
\end{array}$$

and indeed

$$3x^4 - 2x^3 + x^2 + 5x - 4 = (x^2 - x + 6)(3x^2 + x - 2) + 4x + 1.$$

This is a "proof by example" which, while not a proof at all, is highly recommended as a way of convincing yourself of the validity of Theorem 24. Try a couple of your own.

COROLLARY 1. If F is a field, $f(x)$ a polynomial with coefficients in F, and $\alpha \in F$, then $f(\alpha) = 0$ if and only if $f(x) = (x - \alpha)g(x)$ for some polynomial $g(x)$ with coefficients in F.

Proof: If $f(x) = (x - \alpha)g(x)$, then $f(\alpha) = (\alpha - \alpha)g(\alpha) = 0 \cdot g(\alpha) = 0$. On the other hand, we can write $f(x) = q(x)(x - \alpha) + r(x)$ where $r(x) = 0$, or has degree less than 1. But this means that $r(x)$ is a constant. Hence, $f(\alpha) = q(\alpha)(\alpha - \alpha) + r(\alpha) = r(\alpha) = r(x)$, so if $f(\alpha) = 0$, then $r(x) = 0$. Hence, $f(x) = q(x)(x - \alpha)$; setting $g(x) = q(x)$ completes the proof.

COROLLARY 2. If $f(x)$ is a polynomial of degree n with coefficients in a field F, then there are at most n elements $\alpha \in F$ such that $f(\alpha) = 0$.

Proof: Suppose not. Let n be the smallest positive integer for which the theorem fails (it is certainly true for $n = 0$ and $n = 1$; why?). If $f(\alpha) = 0$, then $f(x) = (x - \alpha)g(x)$ by the preceding corollary, and so the degree of $g(x)$ is $n - 1$. Hence, there are at most $n - 1$ elements β of F such that $g(\beta) = 0$. But if $f(\beta) = 0$, then $(\beta - \alpha)g(\beta) = 0$, so either $\beta = \alpha$ or $g(\beta) = 0$. Thus, there are at most n elements β such that $f(\beta) = 0$.

We may use Corollary 2 to obtain information about the group of units in any field. First, we need to know when the direct sum of two groups is cyclic. Let $|G|$ denote the order of the group G.

THEOREM 25. If $G = A \oplus B$, then G is cyclic if and only if A and B are cyclic and $(|A|, |B|) = 1$.

Proof: Suppose G is cyclic and g is a generator of G. Then $g = a + b$ where $a \in A$ and $b \in B$. We show that a is a generator of A. In fact, suppose a_0 is some element of A. Then, since g generates G, we must have $ng = a_0$ for some $n \in Z$. But then $ng = na + nb = a_0$ and, since we have a direct sum, $na = a_0$. Similarly, b generates B. By Theorem 18, the order of g is the least common multiple of the orders of a and b. But, since these all generate their respective groups, this says that $|G|$ is the least common multiple of $|A|$ and $|B|$. On the other hand, $|G| = |A| \cdot |B|$. But this implies that $(|A|, |B|) = 1$. Conversely, if A and B are cyclic with generators a and b of relatively prime orders, then $g = a + b$ has order $|A| \cdot |B| = |G|$, so g generates G.

We can now prove the main theorem.

THEOREM 26. Let F be a finite field and F^* its group of units (that is, F^* consists of the nonzero elements of F). Then F^* is cyclic.

Proof: Let m be the number of elements in F^*. Write $m = q_1 q_2 \cdots q_k$ where $q_1, q_2, \ldots, q_k$ are powers of distinct primes $p_1, p_2, \ldots, p_k$, respectively. Among the elements in F^* whose orders are powers of p_i, choose an element b_i of largest order (possibly $b_i = 1$, the element of order $1 = p_i^0$). Let B_i be the subgroup generated by b_i. We shall show that $F^* = B_1 \oplus B_2 \oplus \cdots \oplus B_k$ and hence, by repeated application of Theorem 25, that F^* is cyclic.

First we observe that B_i consists of *all* elements of F^* whose order is a power of p_i. Indeed, if t_i is the order of b_i, then every element whose order is a power of p_i, since its order cannot exceed t_i, satisfies the equation $x^{t_i} - 1 = 0$. But by Corollary 2, there are at most t_i such elements, and we already have t_i of them in B_i.

We now show that every element in F^* is a product of elements from the B_i. If $g \in F^*$ and the order of g is n, then the subgroup G generated by g is isomorphic to Z_n. But by the corollary to Theorem 20, every element of Z_n is the sum of elements of prime power order (take the a_j to be the prime power factors of n).

Hence, every element of G, in particular g, can be written as a product of elements of prime power order. But the order of any element of F^* divides $m = q_1 q_2 \cdots q_k$, and so any element of prime power order is in one of the B_i.

Finally, we apply another variant of the pigeonhole principle, which says that if you have no more than m pigeons and put them in m pigeonholes in such a way that each pigeonhole gets at least one pigeon, then each pigeonhole gets exactly one pigeon. The pigeons are the strings $(b_1, b_2, \ldots, b_k)$ where $b_i \in B_i$; the pigeonholes are the elements of F^*. We put the pigeon $(b_1, b_2, \ldots, b_k)$ in the pigeonhole $b_1 b_2 \cdots b_k$. Since $|B_i|$ divides $F^* = m = q_1 q_2 \cdots q_k$, and $|B_i|$ is a power of p_i, we have $|B_i| \leq q_i$, so there can be at most m pigeons. The previous paragraph showed that every pigeonhole gets filled, so every hole gets precisely one pigeon—the definition of a direct sum.

COROLLARY. If p is a prime, then U_p is cyclic.

Proof: U_p is the group of units of the finite field Z_p.

4.3 PROBLEMS

1. Show that the set of polynomials with coefficients in a field F forms an integral domain.
2. Which polynomials of Problem 1 are units?
3. Consider the polynomials with coefficients in Q, the field of rational numbers. Show that if $f(x)$ is such a polynomial and $f(\sqrt{2}) = 0$, then $f(x)$ is divisible by $x^2 - 2$ (use the division algorithm).
4. Let F be an infinite field and let $f(x)$ be a polynomial with coefficients in F. Show that if $f(\alpha) = 0$ for all $\alpha \in F$, then $f = 0$.
5. (Remainder theorem) Let $f(x)$ be a polynomial with coefficients in a field F. Show that if $\alpha \in F$ then $f(\alpha)$ is equal to the remainder obtained upon dividing $f(x)$ by $x - \alpha$.

6. Show that the group of units of the field of real numbers is not cyclic. For a bonus, show that in any infinite field the group of units is not cyclic.
7. Show that the group of units of the field of real numbers is a direct sum of a two-element group and the (multiplicative) group of positive real numbers.
8. Show that if x and y are positive integers such that xy is the least common multiple of x and y, then $(x, y) = 1$. Where is this used in this section? Show, more generally, that if z is the least common multiple of x and y then $(x, y) = xy/z$.
9. Write out the final argument for Theorem 26 concerning the repeated application of Theorem 25 to show that F^* is cyclic.
10. Show that, in fact, $b_i \neq 1$ for the b_i considered in the proof of Theorem 26.

4.4 PRIMITIVE ROOTS

In any ring R, an element a is said to be an mth *root* of an element b if $a^m = b$. Let us focus our attention on the mth roots of 1 (mth roots of unity)—that is, elements a such that $a^m = 1$. If R is a field, we know, by the corollary to Theorem 24, that there are at most m mth roots of 1. If R is the rational numbers or the real numbers, there is only one mth root of 1 if m is odd, namely 1 itself, and two mth roots of 1 if m is even, namely 1 and -1. In the complex numbers, there are m mth roots of 1, namely $\cos(2\pi/m) + i\sin(2\pi/m)$, $\cos(4\pi/m) + i\sin(4\pi/m)$, $\cos(6\pi/m) + i\sin(6\pi/m), \ldots, \cos(m2\pi/m) + i\sin(m2\pi/m) = 1$. For example, the four 4th roots of 1 are i, -1, $-i$, and 1.

If ω is an mth root of 1, then so is any power of ω. (Make sure you see why.) If the mth roots of 1 are precisely the powers of ω, we say that ω is a *primitive* mth *root of* 1. Note that i and $-i$ are primitive 4th roots of 1 in the complex numbers whereas 1 and -1 are not. If, in fact, *every* root of 1 is a power of ω, we simply say that ω is a *primitive root of* 1.

In the ring Z_n, every unit is a root of 1, and conversely (why?). So to inquire as to the existence of a primitive root of 1 in Z_n is the same as asking if the group U_n has a generator—this is, if U_n is cyclic. Hence, a generator of U_n is often called a primitive root.

Let's see what n has to be in order that U_n be cyclic. If $n = ab$ and $(a, b) = 1$, then $U_n \cong U_a \oplus U_b$. If $a \neq 2$, then $\phi(a) = |U_a|$ is even since it is divisible by a power of 2 if a is a power of 2, or by $p - 1$ for some odd prime p if a is not a power of 2. Similarly for b. Hence, if $n = ab$ where $(a, b) = 1$ and neither a nor b is 1 or 2 then, by Theorem 25, U_n is not cyclic since $U_n \cong U_a \oplus U_b$ and $(|U_a|, |U_b|) \neq 1$. The only candidates for n which will make U_n cyclic are those numbers that *cannot* be written as ab, where $(a, b) = 1$ and neither a nor b is 1 or 2. What are these numbers?

If n is not a prime power, then it must be divisible by two distinct primes. One of these must be odd; call it p. Then $n = ap^k$, where $(a, p^k) = 1$. If n has the property of the preceding paragraph, then we must have $a = 2$ ($a \neq 1$ since n is not a prime power). Hence, if n is to have this property, n must be a prime power or twice a power of an odd prime. It is not difficult to verify that, conversely, these numbers do indeed have the desired property (do it). Using this observation, we can prove this rather curious result:

THEOREM 27. If U_n is cyclic, then n is either 2, 4, p^k or $2p^k$ where p is an odd prime.

Proof: By the preceding remarks, we need only show that n cannot be $2^3, 2^4, 2^5, \ldots$, that is, that U_{2^k} is not cyclic for $k \geq 3$. Suppose $\bar{a} \in U_{2^k}$. Then a is odd, so $a = 1 + 2b$ for some integer b. Hence, $a^2 = (1 + 2b)^2 = 1 + 4b + 4b^2 = 1 + 4b(b + 1) = 1 + 8c$ for some integer c, since either b or $b + 1$ is even. Now

$$\begin{aligned} a^4 &= (1 + 8c)^2 &&= 1 + 16c + 64c^2 &&= 1 + 16d \\ a^8 &= (1 + 2^4 d)^2 &&= 1 + 2^5 d + 2^8 d^2 &&= 1 + 2^5 e \\ a^{2^4} &= (1 + 2^5 e)^2 &&= 1 + 2^6 e + 2^{10} e^2 &&= 1 + 2^6 f \\ &\ \ \vdots && \ \ \vdots \\ a^{2^j} &= (1 + 2^{j+1} x)^2 = 1 + 2^{j+2} y. \end{aligned}$$

Hence, $a^{2^{k-2}} = 1 + 2^k y \equiv 1 \pmod{2^k}$, so $\bar{a}$ has order less than 2^{k-1} in U_{2^k}. But $|U_{2^k}| = \phi(2^k) = 2^{k-1}$, so $\bar{a}$ cannot generate U_{2^k}. Why does this argument fail for $k = 2$?

We shall now prove the converse of Theorem 27, which is that U_n is cyclic for $n = 2, 4, p^k$, and $2p^k$. Clearly U_2 and U_4 are cyclic. If p is an odd prime, then $U_{2p^k} \cong U_2 \oplus U_{p^k} \cong U_{p^k}$, so if we can show that U_{p^k} is cyclic, we will also know that U_{2p^k} is cyclic. We have already seen that U_p is cyclic (corollary to Theorem 26). To show that U_{p^k} is cyclic, we first prove a technical lemma.

LEMMA. If n is an integer bigger than 1, p is an odd prime, and m is any integer, then

$$\begin{aligned} &(1) \qquad (1 + mp^{n-1})^p \equiv 1 + mp^n \pmod{p^{n+1}}, \\ &(2) \qquad (1 + p)^{p^{n-1}} \equiv 1 + p^n \pmod{p^{n+1}}, \\ &(3) \qquad (1 + p)^{p^n} \equiv 1 \pmod{p^{n+1}}. \end{aligned}$$

Proof: We are really after parts 2 and 3; they say that $1 + p$ has order p^n in $U_{p^{n+1}}$. To prove part 1, we apply the binomial theorem to $(1 + mp^{n-1})^p$ getting

$$1 + pmp^{n-1} + \frac{1}{2}p(p-1)(mp^{n-1})^2 + \cdots + p(mp^{n-1})^{p-1} + (mp^{n-1})^p.$$

Since p is odd, the third term is divisible by $p(p^{n-1})^2 = p^{2n-1}$ and hence, since $n > 1$, is divisible by p^{n+1}. The terms beyond the third term are divisible by $(p^{n-1})^3 = p^{3n-3}$ and so are also divisible by p^{n+1}. Part 2 is true for any positive integer n. It is certainly true for $n = 1$. Suppose it fails to be true for some positive integer, and let n be the least positive integer for which it fails. Then $n > 1$ and it is true for $n - 1$. That is, $(1 + p)^{p^{n-2}} \equiv 1 + p^{n-1} \pmod{p^n}$, so $(1 + p)^{p^{n-2}} = 1 + p^{n-1} + sp^n = 1 + (1 + sp)p^{n-1}$ for some integer s. Thus $(1 + p)^{p^{n-1}} = ((1 + p)^{p^{n-2}})^p = (1 + (1 + sp)p^{n-1})^p$ which, by part 1, is $1 + (1 + sp)p^n \pmod{p^{n+1}}$ which is $1 + p^n \pmod{p^{n+1}}$. Part 3 follows from part 2 upon substituting $n + 1$ for n, yielding $(1 + p)^{p^n} \equiv 1 + p^{n+1} \pmod{p^{n+2}} \equiv 1 \pmod{p^{n+1}}$.

THEOREM 28. If p is an odd prime and k is a positive integer, then U_{p^k} is cyclic.

Proof: For notational convenience, set $k = n + 1$. We must find an element in $U_{p^{n+1}}$ of order $|U_{p^{n+1}}| = p^n(p-1)$. Our candidate is $a^p(1+p)$, where a is an integer that generates U_p. Let t be the order of $a^p(1+p)$ in $U_{p^{n+1}}$. Then t divides $p^n(p-1)$, the order of the group $U_{p^{n+1}}$. We want to show that $t = p^n(p-1)$. Now $(a^p(1+p))^t \equiv 1 \pmod{p^{n+1}} \equiv 1 \pmod{p}$, and so $p-1$ divides t since $a^p(1+p) \equiv a^p \equiv a \pmod{p}$ and a has order $p-1$ in U_p. Thus, since t divides $p^n(p-1)$, t must have the form $p^k(p-1)$. But

$$(a^p(1+p))^{p^{n-1}(p-1)} \equiv (1+p)^{p^{n-1}(p-1)} \not\equiv 1 \bmod (p^{n+1})$$

since, by the lemma, $1 + \mathrm{p}$ has order p^n in $U_{p^{n+1}}$. So t does not divide $p^{n-1}(p-1)$, and so t must equal $p^n(p-1)$.

Notice that one integer, $a^p(1+p)$, serves as a primitive root for *every* power of p.

COROLLARY. U_n is cyclic if and only if n is 2, 4, p^k, or $2p^k$ for p an odd prime.

4.4 PROBLEMS

1. Find all square roots of 1 in Z_8, Z_{15}, and $Z_3 \oplus Z_3$.
2. Show that, in a field, if $\omega^m = 1$ and $\omega^n \neq 1$ for $0 < n < m$, then ω is a primitive mth root of 1. Is this true in a ring (see Problem 1)?
3. Verify directly that $1 + p$ has order p^n in $U_{p^{n+1}}$ for $p = 3$, $n = 2$; $p = 3$, $n = 3$; $p = 5$, $n = 2$.
4. Use the construction in the proof of Theorem 28 to find generators for U_{27}, U_{81}, and U_{125}.
5. Find a generator for U_{59049}.
6. (Structure of U_{2^n}, $n \geq 3$)
   (a) Show that if $x \not\equiv 1 \pmod{2^n}$ and $x \equiv 1 \pmod{4}$, then $x^2 \not\equiv 1 \pmod{2^{n+1}}$.

(b) Observe that $3^2 \not\equiv 1 \pmod{2^4}$. Use (a) to conclude that $3^{2^{n-3}} \not\equiv 1 \pmod{2^n}$ for all $n \geq 4$.

(c) Use (b) to show that 3 has order at least 2^{n-2} in U_{2^n} for $n \geq 3$.

(d) Show that the subgroup generated by 3 in U_{2^n} does not contain -1 for $n \geq 3$, and hence the order of 3 is precisely 2^{n-2} ($3^k \equiv 1$ or $3 \pmod 8$).

(e) Use (c) and (d) to show that $U_{2^n} \cong Z_2 \oplus Z_{2^{n-2}}$ for $n \geq 3$ (multiplicative group isomorphic to additive group).

(f) Use (e) to give another proof that U_{2^n} is not cyclic for $n \geq 3$.

7. Show that a^{p^n} has order $p-1$ in $U_{p^{n+1}}$ (notation as in proof of Theorem 28). Use this and the fact that $1+p$ has order p^n to show that $U_{p^{n+1}} \cong Z_{p-1} \oplus Z_{p^n}$. Use this isomorphism to give another proof of Theorem 28.

CHAPTER V

SUMS OF TWO SQUARES

5.1 THE SQUARE ROOT OF MINUS ONE

The point of departure for this chapter is the question of determining which integers can be written as sums of two squares (of integers). Thus, for a specified integer a, we inquire into the existence of *integer* solutions to the equation

$$x^2 + y^2 = a.$$

For example: $0 = 0^2 + 0^2$, $1 = 1^2 + 0^2$, $2 = 1^2 + 1^2$, $4 = 2^2 + 0^2$, $5 = 2^2 + 1^2$, $8 = 2^2 + 2^2$, $9 = 3^2 + 0^2$, $10 = 3^2 + 1^2$, whereas 3, 6, and 7 cannot be written as sums of two squares.

Notice that if a is a sum of two squares, then so is ab^2 for any integer b. Indeed, if $a = x^2 + y^2$, then $ab^2 = x^2b^2 + y^2b^2 = (xb)^2 + (yb)^2$. For example, since we know that 10 is the sum of two squares, then so is $90 = 10 \cdot 3^2$; in fact, $90 = 9^2 + 3^2$. Similarly, $160 = 10 \cdot 4^2 = 12^2 + 4^2$. On the other hand, if $c = s^2 + t^2$ and b is a common factor of s and t, then $c = s^2 + t^2 = (xb)^2 + (yb)^2 = (x^2 + y^2)b^2$, so $c = ab^2$ where $a = x^2 + y^2$ is the sum of two squares; and by taking $b = (s, t)$, we can arrange to have $(x, y) = 1$. Thus, an integer is a sum of two squares if and only if it can be written as ab^2 where a is a sum of two *relatively prime* squares.

Suppose a is the sum of two relatively prime squares—that is, $a = x^2 + y^2$ with $(x, y) = 1$. What can we say about the primes in the factorization of a? If $p \mid a$, then $x^2 + y^2 \equiv 0 \pmod{p}$. Since $(x, y) = 1$, p cannot divide both x and y and hence, since $p \mid (x^2 + y^2)$, p cannot divide either x or y. Thus, in the field Z_p, we have $x^2 = -y^2$, and we can divide this equation by y^2 getting $(x/y)^2 = -1$. This says that there is an element in Z_p whose square is -1.

What primes p have the property that there is a square root of -1 in Z_p? The most important way to classify primes is into even primes and odd primes. Of course, 2 is the only even prime, but its behavior is peculiar enough to single it out from the others. The most important distinction among the odd primes turns out to be their residues modulo 4. If p is an odd prime, then either $p \equiv 1 \pmod{4}$ or $p \equiv 3 \pmod{4}$ (why?). In fact, this is true of any odd number, prime or not. We now state the facts about when there is a square root of -1 in Z_p.

THEOREM 29. If p is a positive prime, then there is an element $x \in Z_p$ such that $x^2 = -1$ if and only if $p = 2$ or $p \equiv 1 \pmod{4}$.

Proof: If $p = 2$, then $1 = -1$, so $1^2 = 1 = -1$. In any field where $1 \neq -1$, an element x has the property that $x^2 = -1$ if and only if x has order 4 in the group of units of the field. To see this, we note that if $x^2 = -1$, then the first four powers of x are x, -1, $-x$, 1, so x has order 4. On the other hand, if x has order 4, then $(x^2)^2 = x^4 = 1$, so $(x^2)^2 - 1 = 0$, so $(x^2 - 1)(x^2 + 1) = 0$. Since we are in a field, this means that either $x^2 - 1 = 0$ or $x^2 + 1 = 0$ —either $x^2 = 1$ or $x^2 = -1$. The former would imply that x has order 2; hence, the latter must hold.

If $p \neq 2$, then Z_p is a field where $1 \neq -1$, so the existence of an element x such that $x^2 = -1$ is equivalent to the existence of an element of order 4 in U_p. Now $|U_p| = p - 1$, and since the order of any element divides the order of the group, if U_p is to have an element of order 4, we must have $4 \mid (p - 1)$, or $p \equiv 1 \pmod 4$. Conversely, if $p \equiv 1 \pmod 4$, then $p = 4k + 1$ and U_p is a cyclic group of order $p - 1 = 4k$. If g is a generator of U_p, then g has order $4k$, so g^k has order 4.

COROLLARY. If an integer c can be expressed as the sum of two relatively prime squares, then c is not divisible by any prime $p \equiv 3 \pmod 4$.

5.1 PROBLEMS

1. Find the square roots of minus one in the fields Z_5, Z_{13}, Z_{17}, Z_{29}, and Z_{37}.
2. Prove the corollary to Theorem 29.
3. Find all positive integers less than 60 that are the sums of two squares. Write each one in the form ab^2 where a is a sum of two relatively prime squares. Note which ones can be written as the sum of two squares in more than one way.
4. What real numbers can be written as sums of two squares (of real numbers)? What complex numbers can be written as sums of two squares (of complex numbers)?
5. Can an integer be the sum of two squares of noninteger rational numbers?

5.2 GAUSSIAN INTEGERS

We saw in the last section that if a is a sum of two squares and b is a square, then ab is a sum of two squares. More generally, it is true that if a and b are both sums of two squares, then so is ab. We could verify this by simply writing down the relevant identity, but it is more enlightening to introduce a notion that leads naturally to that identity.

A *Gaussian number* is a complex number of the form $x + yi$ where x and y are *rational* numbers (and $i^2 = -1$). Clearly the sum, difference, and product of two Gaussian numbers are again Gaussian numbers. Since the complex numbers form an integral domain, the same is true of the Gaussian numbers; the required properties all follow from the fact that they hold for the complex numbers. Indeed, the Gaussian numbers form a field, as you can see by verifying that the inverse of $x + yi$ is $x/(x^2 + y^2) + (-y/(x^2 + y^2))i$, provided that $x + yi \neq 0$. If $\alpha = x + yi$ is a Gaussian number, we define the *norm* of α by $N(\alpha) = (x + yi)(x - yi) = x^2 + y^2$. If you are familiar with the elementary geometry of complex numbers you will observe that $N(\alpha)$ is the square of the distance from α to 0 in the complex plane, and hence $N(\alpha\beta) = N(\alpha)N(\beta)$. We exhibit the computations.

THEOREM 30. If α and β are Gaussian numbers, then $N(\alpha\beta) = N(\alpha)N(\beta)$.

Proof: Let $\alpha = x + yi$ and $\beta = s + ti$. Then $\alpha\beta = (x + yi)(s + ti) = (xs - yt) + (xt + ys)i$. Hence,

$$N(\alpha\beta) = (xs - yt)^2 + (xt + ys)^2 = x^2s^2 - 2xyst + y^2t^2 + x^2t^2 + 2xyst + y^2s^2 = x^2s^2 + y^2t^2 + x^2t^2 + y^2s^2 = (x^2 + y^2)(s^2 + t^2) = N(\alpha)N(\beta).$$

A *Gaussian integer* is a Gaussian number $x + yi$ where x and y are *integers*. Notice that an integer is the sum of two squares if and only if it is the norm of some Gaussian integer. With this in mind, we may prove the statement at the beginning of this section.

COROLLARY. If a and b are sums of two squares, then so is ab. If $a = x^2 + y^2$ and $b = s^2 + t^2$, then $ab = (xs - yt)^2 + (xt + ys)^2$.

Proof: If a and b are sums of two squares, then $a = N(x + yi)$ and $b = N(s + ti)$ for some integers x, y, s, and t. Hence, $ab = N((x + yi)(s + ti)) = N((xs - yt) + (xt + ys)i) = (xs - yt)^2 + (xt + ys)^2$.

The Gaussian integers form an integral domain (why?). Hence, it is natural to inquire into the possibility of unique factorization into primes. The norm plays the role here that absolute value did for the integers and degree did for polynomials over a field. We must pay a little closer attention to units and associates than we did for the integers, because we don't have a nice notion of positiveness. An integer is a unit if and only if its absolute value is 1; a polynomial is a unit if and only if its degree is 0. Similarly, you can tell when a Gaussian integer is a unit by looking at its norm.

THEOREM 31. A Gaussian integer α is a unit if and only if $N(\alpha) = 1$. This occurs precisely when $\alpha = 1, -1, i$, or $-i$.

Proof: If α is a unit, then $\alpha\beta = 1$ for some β. Hence, $N(\alpha\beta) = N(1) = 1^2 + 0^2 = 1$. Thus, $N(\alpha)N(\beta) = 1$. But $N(\alpha)$ and $N(\beta)$ are both nonnegative integers, so $N(\alpha) = 1$.

If $N(\alpha) = 1$, then $x^2 + y^2 = 1$ where $\alpha = x + yi$. A little thought about the size of $x^2 + y^2$ when x and y are integers will convince you that this can only happen if $x = \pm 1$ and $y = 0$ or if $x = 0$ and $y = \pm 1$—that is, if $\alpha = 1, -1, i$ or $-i$.

Finally, the numbers $1, -1, i$, and $-i$ are units since $1 \cdot 1 = 1$, $(-1)(-1) = 1$, and $i(-i) = 1$.

Notice the structure of the proof of Theorem 31. We wished to prove that the three statements

(1) α is a unit
(2) $N(\alpha) = 1$
(3) $\alpha = 1, -1, i$, or $-i$

all said the same thing. To do this, we showed that (2) followed from (1), that (3) followed from (2), and that (1) followed from (3). Hence, any one follows from any other. This type of circular chain of implication often saves time when it is desired to show that three or more statements are equivalent.

If α is a Gaussian integer, then α can always be written as a product: $\alpha = 1 \cdot \alpha$, $\alpha = (-1)(-\alpha)$, $\alpha = i(-i\alpha)$, and $\alpha = (-i)(i\alpha)$. However, these are all uninteresting factorizations because one of the factors is a unit.

DEFINITION. If π is a Gaussian integer that is not a unit, such that whenever $\pi = \alpha\beta$ for Gaussian integers α and β then either α or β is a unit, then π is called a *Gaussian prime*.

Compare this definition to the definition of prime for the integers, remembering that the units of Z are precisely 1 and -1. Just like for the integers, we can express every Gaussian integer as a product of primes and units. Specifically, we have the following corollary to Theorem 31.

COROLLARY. Every nonzero Gaussian integer is either a unit, a prime, or a product of primes.

Proof: Suppose not. Let α be a nonzero Gaussian integer of smallest norm which is neither a unit, a prime, nor a product of primes. Since α is not a unit or a prime, then $\alpha = \beta\gamma$ where neither β nor γ is a unit. Hence, by Theorem 31, $N(\beta)$ and $N(\gamma)$ are bigger than 1, so, by Theorem 30, they are smaller than $N(\alpha)$. By the minimality of $N(\alpha)$, β and γ must be primes or products of primes. Thus α is a product of primes; but α was chosen *not* to be a product of primes—a contradiction.

The underlying idea in this proof is to take a Gaussian integer and factor it as much as you can. Since every nontrivial factorization decreases norms, this procedure eventually stops, and you are left with Gaussian primes. Compare this proof with the proof of Theorem 2.

Divisibility is defined for Gaussian integers (or for any integral domain) just like for integers: we say that $\alpha \mid \beta$ if $\beta = \alpha\gamma$ for some Gaussian integer γ. If α does not divide β, we can still divide β by α and get a small remainder; we have a division algorithm.

DIVISION ALGORITHM FOR GAUSSIAN INTEGERS. If β and α are Gaussian integers, and $\alpha \neq 0$, then there exist Gaussian integers γ and ρ such that $\beta = \gamma\alpha + \rho$ and $N(\rho) < N(\alpha)$.

Proof: Write $\beta/\alpha = x + yi$, where x and y are rational numbers. Choose integers s and t such that $|x - s|$ and $|y - t|$ are no greater than 1/2. Let $\gamma = s + ti$. Then γ is a Gaussian integer and $N(\beta/\alpha - \gamma) = (x-s)^2 + (y-t)^2 \leq 1/4 + 1/4 = 1/2$. Let $\rho = \beta - \gamma\alpha = \alpha(\beta/\alpha - \gamma)$. Then $\beta = \gamma\alpha + \rho$ and $N(\rho) = N(\alpha)N(\beta/\alpha - \gamma) \leq N(\alpha) \cdot 1/2 < N(\alpha)$.

As an illustration, consider the numbers $\beta = 17 - 4i$ and $\alpha = 3 - 2i$. Then $1/\alpha = (3 + 2i)/13$, so $\beta/\alpha = (59 + 22i)/13 = 59/13 + (22/13)i = x + yi$. By choosing $s = 5$ and $t = 2$, we have $|x-s| = |59/13 - 5| = 6/13 \leq 1/2$ and $|y-t| = |22/13 - 2| = 4/13 \leq 1/2$. Letting $\gamma = 5 + 2i$, we have $\gamma\alpha = (5 + 2i)(3 - 2i) = 19 - 4i$, so $\beta = 17 - 4i = (19 - 4i) - 2 = \gamma\alpha - 2$. Thus, $\beta = \gamma\alpha + \rho$ where $\rho = -2$ and $N(\rho) = 4 < 13 = N(\alpha)$.

The notions of "ideal" and "principal ideal" also generalize immediately to any ring R; simply replace "Z" by "R" in the definition. A subset I of R is an *ideal* if

I_1. $0 \in I$.
I_2. If $x \in I$ and $y \in I$, then $x + y \in I$.
I_3. If $x \in I$ and $z \in R$, then $zx \in I$.

Similarly, if $m \in R$, then the set of all elements of the form rm, for $r \in R$, is called the *principal ideal generated by m*. A principal ideal is an ideal (why?); the nice thing about the Gaussian integers is that the converse is true: every ideal is a principal ideal. This is proved, just like for Z, by using the division algorithm.

THEOREM 32. Every ideal in the Gaussian integers is principal.

Proof: Let I be an ideal. If $I = \{0\}$, then I consists of all the multiples of 0 and so is principal. If $I \neq \{0\}$, choose a nonzero element $\alpha \in I$ of smallest norm. Since I satisfies property I_3, every multiple of α is in I. We must show, conversely, that every element of I is a multiple of α. Let β be an element of I. Then $\beta = \gamma\alpha + \rho$, where $N(\rho) < N(\alpha)$. But then $\rho = \beta + (-\gamma)\alpha$. Now $\beta \in I$ by assumption; $(-\gamma)\alpha \in I$ by I_3; thus $\beta + (-\gamma)\alpha \in I$ by I_2. But this says that $\rho \in I$. Since $N(\rho) < N(\alpha)$, if $\rho \neq 0$ we have a contradiction since α was chosen as a nonzero element of I of smallest norm. Hence $\rho = 0$, so $\beta = \gamma\alpha$ is a multiple of α.

Notice that the proof is practically word for word the same as the proof of Theorem 3. What we are really proving in both cases is that a suitable division algorithm implies that every ideal is principal. This property is all we need to prove unique factorization into primes, as we did for Z. We shall content ourselves here with proving a key corollary, leaving the proof of unique factorization as an exercise.

COROLLARY. If π is a Gaussian prime and $\pi \mid \alpha\beta$, then $\pi \mid \alpha$ or $\pi \mid \beta$.

Proof: Consider the set I of Gaussian integers of the form $\sigma\alpha + \tau\pi$, where σ and τ are Gaussian integers. It is easily verified that I is an ideal. By Theorem 32, I is the set of multiples of some Gaussian integer δ. Since $\pi = 0 \cdot \alpha + 1 \cdot \pi \in I$, we have $\pi = \varepsilon\delta$ for some Gaussian integer ε. Since π is a Gaussian prime, either ε or δ must be a unit. If ε is a unit, then $\delta = \varepsilon^{-1}\pi$, so $\pi \mid \delta$. But $\alpha = 1 \cdot \alpha + 0 \cdot \pi \in I$, so $\delta \mid \alpha$. Hence, if ε is a unit, then $\pi \mid \alpha$. On the other hand, if δ is a unit, then, since $\delta \in I$, we have $\delta = \sigma\alpha + \tau\pi$ for some Gaussian integers σ and τ. Multiplying this equation by $\delta^{-1}\beta$, we get $\beta = (\sigma\delta^{-1})\alpha\beta + (\tau\delta^{-1}\beta)\pi$. But $\pi \mid \alpha\beta$, and everything in sight is a Gaussian integer, so $\pi \mid \beta$.

5.2 PROBLEMS

1. Show that $N(x) = x^2$ for any integer x.
2. Show that if α and β are Gaussian numbers and $\beta \neq 0$, then $N(\alpha/\beta) = N(\alpha)/N(\beta)$.
3. Find all Gaussian integers α such that $N(\alpha) = 2$.
4. Show that a Gaussian integer $x + iy$ is divisible by $1 + i$ if and only if $x + y$ is even.
5. Show that a Gaussian integer $x + yi$ is divisible by an integer n if and only if $n \mid x$ and $n \mid y$. Is such a statement true if n is a Gaussian integer?
6. Show that if α and u are Gaussian integers, and u is a unit, then α is a Gaussian prime if and only if $u\alpha$ is a Gaussian prime.
7. Show that if π is a Gaussian prime and α is a Gaussian integer, then either $\pi \mid \alpha$ or the only common factors of π and α are units.

8. Let α and β be Gaussian integers whose only common factors are units. Show that there exist Gaussian integers σ and τ such that $\sigma\alpha + \tau\beta = 1$.

9. In any integral domain, two elements a and b are said to be *associates* if $a \mid b$ and $b \mid a$. Show that a and b are associates if and only if $a = ub$ for some unit u. Observe that this agrees with what we know for Z.

10. What are the associates of the Gaussian integer $2 + 3i$?

11. Find Gaussian integers γ and ρ such that $\beta = \gamma\alpha + \rho$ and $N(\rho) < N(\alpha)$ where: $\alpha = 3$, $\beta = 7$; $\alpha = 3 - 2i$, $\beta = 1 + i$; $\alpha = 1 + i$, $\beta = 3 - 2i$; $\alpha = 2 + i$, $\beta = 6$.

12. (The fundamental theorem of arithmetic for Gaussian integers.) Show that if $\alpha_1, \alpha_2, \ldots, \alpha_n$ and $\beta_1, \beta_2, \ldots, \beta_m$ are Gaussian primes such that $\alpha_1 \alpha_2 \cdots \alpha_n = \beta_1 \beta_2 \cdots \beta_m$, then $n = m$ and, after possibly relabeling the β's, $\beta_j = u_j\alpha_j$, where u_j is a unit, $j = 1, 2, \ldots, n$. (Use the corollary to Theorem 32 to show that $\alpha_1 \mid \beta_j$ for some j; relabel so $j = 1$; use the fact that β_1 is prime to write it in the desired form; cancel α_1 from both sides; repeat on $\alpha_2 \cdots \alpha_n = u_1 \beta_2 \cdots \beta_m$, etc.)

5.3 GAUSSIAN PRIMES

We may use the results of the preceding section to determine precisely when a Gaussian integer is a Gaussian prime and when an integer is the sum of two squares. We first determine which integers are Gaussian primes—that is, which integers admit only trivial factorizations even if we allow the factors to be Gaussian integers. Certainly such integers must be primes, but that is not sufficient. The integer 2 is a prime, but since $2 = (1+i)(1-i)$, it is not a Gaussian prime. Similarly,

$$13 = (2+3i)(2-3i)$$

is not a Gaussian prime. When can a prime p be written as the product of two Gaussian integers in a nontrivial way?

THEOREM 33. Let p be a positive prime. Then the following statements are equivalent:

(1) $p \equiv 3 \pmod 4$.
(2) p cannot be written as the sum of two squares.
(3) p is a Gaussian prime.

Proof: If $p \equiv 3 \pmod 4$, then p cannot be written as the sum of two squares since any square is equivalent to 0 or 1 modulo 4, so the sum of two squares cannot be equivalent to 3 modulo 4.

Suppose p cannot be written as the sum of two squares; we wish to show that p is a Gaussian prime. Let $p = (a+bi)(c+di)$. Then $p^2 = N(p) = N((a+bi)(c+di)) = N(a+bi)\,N(c+di)$. Thus, either $N(a+bi) = p$, $N(a+bi) = 1$, or $N(c+di) = 1$. If $N(a+bi) = p$, then $p = a^2 + b^2$, contrary to our assumption that p cannot be written as the sum of two squares. Thus, either $N(a+bi) = 1$ or $N(c+di) = 1$, so either $a+bi$ or $c+di$ is a unit—that is, p is a Gaussian prime.

Finally suppose p is a Gaussian prime; we wish to show that $p \equiv 3 \pmod 4$. If not, by Theorem 29 we can find an integer x such that $x^2 \equiv -1 \pmod p$, that is, $p \mid (x^2 + 1)$. But $x^2 + 1 = (x+i)(x-i)$ and since $p \mid (x^2 + 1)$, then either $p \mid (x+i)$ or $p \mid (x-i)$. This is clearly impossible, for if $x \pm i = p(a+bi)$, then $x = pa$ and $\pm 1 = pb$, and so $p \mid 1$, which is absurd. Thus, $p \equiv 3 \pmod 4$.

COROLLARY. A positive integer a is the sum of two squares if and only if $a = b^2c$ where c is not divisible by any positive prime $p \equiv 3 \pmod 4$.

Proof: Suppose $a = s^2 + t^2$ and let $(s, t) = b$. Then $a = (bx)^2 + (by)^2 = b^2(x^2 + y^2)$ where $(x, y) = 1$. Letting $c = x^2 + y^2$, we have $a = b^2c$ where c is the sum of two relatively prime squares. Hence, by the corollary to Theorem 29, c is not divisible by any prime $p \equiv 3 \pmod 4$.

Conversely, if c is not divisible by any prime $p \equiv 3 \pmod 4$, then c is a product of primes each of which, by Theorem 33, is a sum of two squares. Hence, by the corollary to Theorem 30, b^2c is the sum of two squares.

You should convince yourself that this corollary says that a positive integer a is the sum of two squares if and only if, upon factoring

a into positive primes, each prime $p \equiv 3 \pmod 4$ appears an even number of times.

We turn now to an explicit description of the Gaussian primes.

THEOREM 34. A Gaussian integer α is a Gaussian prime if and only if either

(1) $N(\alpha)$ is a prime.

or (2) $\alpha = \pm p$ or $\alpha = \pm pi$, where p is a positive prime congruent to 3 modulo 4.

Proof: If $\alpha = \pm a$ or $\alpha = \pm ai$, where a is a positive integer, then α will be a Gaussian prime if and only if a is. But, by Theorem 33, this means that (2) holds. Otherwise $\alpha = a + bi$ where neither a nor b is 0. If $N(\alpha)$ is a prime, then if $\alpha = \beta\gamma$ we have $N(\alpha) = N(\beta)N(\gamma)$, so either $N(\beta)$ or $N(\gamma)$ must be 1; hence, either β or γ is a unit. Thus, if (1) holds then α is a Gaussian prime. Conversely, if $\alpha = a + bi$ where neither a nor b is 0, and α is a Gaussian prime, then if $N(\alpha) = \alpha(a-bi) = cd$, either $\alpha \mid c$ or $\alpha \mid d$. Suppose $c = \alpha\gamma$. Then $\alpha(a-bi) = \alpha\gamma d$, so $a-bi = \gamma d$. Hence, $d \mid a$ and $d \mid b$, so $d \mid \alpha$ since $\alpha = a + bi$. Since neither a nor b is 0, d must be ± 1. The same argument shows that if $d = \alpha\gamma$, then $c = \pm 1$. Thus if $N(\alpha) = cd$, then either $c = \pm 1$ or $d = \pm 1$—that is, $N(\alpha)$ is a prime.

5.3 PROBLEMS

1. Show that if n is an integer such that $n \equiv 3 \pmod 4$, then n cannot be written as the sum of two squares.
2. Find a positive integer n such that $n \equiv 1 \pmod 4$ but n cannot be written as the sum of two squares. Can such an n be prime?
3. Write the following Gaussian integers as products of primes: $3 - i$, $4 + 7i$, $5 + i$. (One way to find factors is to use the fact that any factor of α is a factor of $N(\alpha)$—proof?)

4. Show that if an integer can be written as the sum of two squares of *rational numbers*, then it can be written as the sum of two squares of *integers*.
5. Show that a positive prime p can be written as the sum of two squares in at most one way (if $x^2 + y^2 = u^2 + v^2 = p$ then $(x+iy)(x-iy) = (u+iv)(u-iv)$ and everything in sight is a Gaussian prime).
6. What rational numbers can be written as the sum of two squares of rational numbers? (Note that r is such a number if and only if $r = N(\alpha)$ for some Gaussian number α.) Which of the following rational numbers can be so written: 1, 1/2, −1/2, 5/6, 14/343, 18/605?

5.4 PYTHAGOREAN TRIPLES

The lengths of the sides of a right triangle are related by the equation

$$x^2 + y^2 = z^2. \tag{1}$$

A problem that has its roots in antiquity concerns the possibility of choosing a unit of length such that x, y, and z are *integers*. At first, it seemed obvious that this could be done. Then someone (some say Pythagoras) observed that if $x = y$, then $(z/x)^2 = 2$. (Note that these statements are independent of the choice of unit of length.) But it could be shown that 2 is not the square of any rational number, so there could be no unit of length that made both x and z integers. On the other hand, there were right triangles whose sides had integer lengths, like 3, 4, 5. The question is how many such triangles are there, and what are they?

Algebraically, we seek integer solutions to equation 1. Since we are interested in honest triangles, we want x, y, and z to be positive (anyway, the negative cases follow trivially from the positive ones since everything is squared; and if anything is zero there is no problem). A triple x, y, z of positive integers satisfying $x^2 + y^2 = z^2$ is called a *Pythagorean triple*. Each Pythagorean triple corresponds to a right triangle whose sides have integer lengths.

If x, y, z is a Pythagorean triple, then so is nx, ny, nz for any positive integer n. Algebraically, this says that if $x^2 + y^2 = z^2$, then

$(nx)^2 + (ny)^2 = n^2(x^2 + y^2) = n^2z^2 = (nz)^2$. Geometrically, it amounts to taking a unit of length that is $1/n$ as long as the original (so the side that was x units long is now nx units long). Either way of looking at it, we don't get anything very different. We say that a Pythagorean triple x, y, z is *primitive* if $(x, y) = 1$. Notice that this also implies that $(x, z) = (y, z) = 1$.

5.4 PROBLEMS

1. Show that any Pythagorean triple can be written in the form nx, ny, nz where x, y, z is a primitive Pythagorean triple and n is a positive integer (If $a^2 + b^2 = c^2$ and $(a, b) = n$, let $a = nx$ and $b = ny$. Show that $c = nz$ and that x, y, z is a primitive Pythagorean triple).

   In light of this, we may content ourselves with finding the primitive Pythagorean triples. Two examples are 3, 4, 5 and 5, 12, 13. Before proceeding, you should try to find another by trial and error.

   Now $x^2 + y^2$ is simply $N(x + iy)$, so we are really looking for Gaussian integers whose norms are squares (of integers). An easy way to get Gaussian integers with square norms is to take squares of Gaussian integers. Indeed, $N(\alpha^2) = N(\alpha)^2$ is a square. Thus, good candidates are $x + iy = (u + iv)^2$, or, $x = u^2 - v^2$ and $y = 2uv$ for some integers u and v. Notice that this makes y even; but this is no loss because if $x^2 + y^2 = z^2$, then x and y cannot both be odd (look at this equation modulo 4) so we might as well take y to be even. However, just any old u and v won't do; we want x and y to be positive and $(x, y) = 1$.

2. Show that if u and v are integers such that $u > v > 0$, $(u, v) = 1$, and either u or v is even, then $x + iy = (u + iv)^2$ and $z = N(u + iv)$ defines a primitive Pythagorean triple x, y, z. (The only real difficulty is showing that $(x, y) = 1$. Show that if p is a prime dividing $y = 2uv$, then p does not divide x; remember that $(u, v) = 1$.)

3. Find six primitive Pythagorean triples and verify directly that they satisfy the definition.

We now have a machine that produces primitive Pythagorean triples in abundance. The question is whether we get them *all* this way. The big problem is to show that if x, y, z is a primitive Pythagorean triple with y even, then $x + iy$ is a square (of a Gaussian integer). We approach this by factoring $x + iy$ into Gaussian primes. It will be convenient to have the following notation and terminology: if $\alpha = a + bi$, then $\bar{\alpha} = a - bi$ is the *conjugate* of α. Note that $N(\alpha) = N(\bar{\alpha}) = \alpha\bar{\alpha}$.

4. Show that if π is a Gaussian prime, then so is $\bar{\pi}$ (use Theorem 34).

5. Let $x, y \in Z$ such that $(x, y) = 1$. Observe that if p is a prime, then p does not divide $x + iy$.
   (a) Show that if π is a Gaussian prime factor of $x + iy$, then $N(\pi)$ is a prime (use Theorem 34).
   (b) Let $x + iy = \pi_1\pi_2 \cdots \pi_t$ where the π's are Gaussian primes. Show that if $N(\pi_i) = N(\pi_j) = p$, then $\pi_j = s\pi_i$ for some Gaussian unit s (apply the lemma following Theorem 32 to $\pi_i\bar{\pi}_i = \pi_j\bar{\pi}_j$; if $\pi_i \mid \bar{\pi}_j$, then $\bar{\pi}_j \mid \pi_i$, so $p \mid (x + iy)$, which is impossible).

6. Let x, y, z be a primitive Pythagorean triple with y even and $x + iy = \pi_1\pi_2 \cdots \pi_t$ where the π's are Gaussian primes.
   (a) Show that among the primes $N(\pi_i)$, $i = 1, 2, \ldots, t$, each prime occurs an even number of times ($N(x + iy) = z^2$).
   (b) Conclude that $x + iy$ may be written, possibly after some relabeling, as $s(\pi_1 \pi_2 \cdots \pi_r)^2$, where s is a Gaussian unit.
   (c) Use the fact that x is odd to show that the s in (b) is ± 1. Use the fact that -1 is a square to show that $x + iy = (u + iv)^2$ for some integers u and v.
   (d) Show that x, y, z is of the form described in Problem 2.

CHAPTER VI

QUADRATIC RESIDUES

6.1 SQUARES AND NONSQUARES IN Z_p

We observed in the proof of Theorem 33 that a sum of two squares cannot be congruent to 3 modulo 4. The reason was that the only squares in Z_4 are 0 and 1. Questions concerning squares in Z_n have played a central role in the history of number theory.

Let $n > 1$ be an integer. We separate the integers a such that $(a, n) = 1$ into two classes. An integer is said to be a *quadratic residue* of n if there is an integer x such that $x^2 \equiv a \pmod{n}$. Otherwise a is called a *quadratic nonresidue* of n. Observe that this property depends only on the residue class of a modulo n and is equivalent to separating U_n into two classes: the squares and the nonsquares. Hence we shall often identify a with its residue class modulo n. In U_4, for example, 1 is a quadratic residue while 3 is not.

First we look at the most important case, where n is an odd prime.

THEOREM 35. If p is an odd positive prime and r is a primitive root in Z_p (that is, a generator of U_p), then the quadratic residues of p correspond to even powers of r; the quadratic nonresidues correspond to odd powers of r. Furthermore,

(1) There are precisely $(p-1)/2$ quadratic residues of p and $(p-1)/2$ quadratic nonresidues.
(2) The product of two residues or of two nonresidues is a residue. The product of a residue and a nonresidue is a nonresidue.

Proof: Certainly the even powers of r are squares since $r^{2k} = (r^k)^2$. Conversely suppose r^{2k+1} is a square. Then $r^{2k+1} = (r^s)^2$ for some s, since r is a primitive root. Hence $r^{2k+1} = r^{2s}$, so $r^{2(k-s)+1} = 1$. But the order of r is $p-1$, which is even, and hence cannot divide $2(k-s)+1$, which is odd. Thus, the quadratic residues are precisely $r^2, r^4, \ldots, r^{p-1} = 1$, while the nonresidues are $r, r^3, \ldots, r^{p-2}$. Count them to prove (1); (2) is clear.

In order to facilitate statements about quadratic residues, we introduce the *Legendre symbol*, (a/p), defined for integers a and odd positive primes p such that $(a, p) = 1$ by

$$(a/p) = 1 \quad \text{if } a \text{ is a quadratic residue of } p$$
$$(a/p) = -1 \quad \text{if } a \text{ is a quadratic nonresidue of } p.$$

This definition is motivated by part 2 of Theorem 35, which can be rewritten, using the Legendre symbol, to read

$$(ab/p) = (a/p)(b/p). \tag{1}$$

If we fix p and view the Legendre symbol as a function of its first variable, it may be considered as a function from the group U_p to the two-element group $\{1, -1\}$. Equation 1 is then just like condition 3 in the definition of an isomorphism. A function from one group to another that satisfies condition 3 is called a *homomorphism*. It may or may not be an isomorphism. The Legendre symbol fails to be $1-1$ in general; $1 \not\equiv 4 \pmod 5$, but $(1/5) = (4/5) = 1$.

THEOREM 36. If p is an odd positive prime and a and b are relatively prime to p, then

(1) $(ab/p) = (a/p)(b/p)$.
(2) $(a/p) \equiv a^{(p-1)/2} \pmod p$;
in particular $(-1/p) = (-1)^{(p-1)/2}$.

Proof: Part 1 is just part 2 of Theorem 35. To prove part 2, let r be a generator of U_p. Then $r^{(p-1)/2} = -1$ since $0 = r^{p-1} - 1 = (r^{(p-1)/2} - 1)(r^{(p-1)/2} + 1)$ and $r^{(p-1)/2} - 1 \neq 0$ since r has order $p-1$. Thus, if $a = r^s$ then $a^{(p-1)/2} = (-1)^s = (a/p)$.

6.1 PROBLEMS

1. Find the quadratic residues of 3, 5, 7, 11, 13, and 17.
2. For what primes is -1 a quadratic residue? For what primes is 4 a quadratic residue?
3. Show that the quadratic residues of an odd prime p form a subgroup of U_p. What are the cosets of this subgroup?
4. Let p be an odd prime. Show that the polynomial $x^2 + bx + c$ with integer coefficients has two roots in Z_p if $b^2 - 4c$ is a quadratic residue of p, one root if $b^2 \equiv 4c \pmod p$, and no roots otherwise ("complete the square" by adding and subtracting $b^2/4$).
5. Prove Theorem 35 for p^n instead of p, replacing $(p-1)/2$ by $p^{n-1}(p-1)/2$.
6. Use Problem 5 to show that, for p an odd prime, y is a quadratic residue of p^n if and only if y is a quadratic residue of p.
7. Use Problem 6 in section 4.4 to show that every element of U_{2^n}, $n \geq 3$, can be written uniquely as $\pm 3^k$, $1 \leq k \leq 2^{n-2}$. Conclude that x is a quadratic residue of 2^n, $n \geq 3$, if and only if $x \equiv 1 \pmod 8$.

8. Show that x is a quadratic residue of n if and only if x is a quadratic residue of every prime power dividing n.
9. Show that the function assigning to each integer its residue class modulo n is a homomorphism from the group Z to the group Z_n. Let F be a field and $\alpha \in F$. Show that the function assigning to each polynomial $f(x)$ with coefficients in F the element $f(\alpha) \in F$ is a homomorphism from the (additive) group of polynomials with coefficients in F to the (additive) group F.

6.2 GAUSS'S LEMMA

We turn now to a key technical lemma concerning (a/p).

LEMMA (Gauss). Let a be an integer and p an odd positive prime not dividing a. Consider the elements

$$a, 2a, 3a, \ldots, \frac{1}{2}(p-1)a$$

of U_p. Let k be the number of these elements that, when considered as elements of the set $\{1, 2, \ldots, p-1\}$, are greater than $p/2$. Then

$$(a/p) = (-1)^k.$$

Proof: Note that the elements $a, 2a, 3a, \ldots, ((p-1)/2)\,a$ are distinct and none is the negative of another (why?). Denote by $a_1, a_2, \ldots, a_k$ those that are greater than $p/2$ and by $a_{k+1}, a_{k+2}, \ldots, a_{(p-1)/2}$ those that are less than $p/2$. Then the numbers $p-a_1, p-a_2, \ldots, p-a_k, a_{k+1}, a_{k+2}, \ldots, a_{(p-1)/2}$ are all less than $p/2$, and are distinct. Hence, they are simply the numbers 1, 2, $\ldots$, $(p-1)/2$ rearranged (the pigeonhole principle). Thus,

$$\begin{aligned} 1 \cdot 2 \cdots \frac{1}{2}(p-1) &\equiv (-a_1)(-a_2)\cdots(-a_k)\, a_{k+1}a_{k+2}\cdots a_{(p-1)/2} \\ &\equiv (-1)^k 1 \cdot 2 \cdots \frac{1}{2}(p-1)a^{(p-1)/2} \pmod{p}, \end{aligned}$$

and so $(-1)^k \equiv a^{(p-1)/2} \equiv (a/p) \pmod{p}$.

A useful corollary to Gauss's lemma makes use of the *greatest integer function* $[x]$ defined for real numbers x by

$$[x] = \text{greatest integer not exceeding } x.$$

For example: $\left[3\frac{1}{2}\right] = 3$, $[7] = 7$, $\left[-3\frac{1}{2}\right] = -4$.

COROLLARY. Let a be an integer and p an odd positive prime not dividing a. Let $M = [a/p] + [2a/p] + \cdots + \left[\frac{1}{2}(p-1)a/p\right]$.

Then

(1) If a is odd, $(a/p) = (-1)^M$.
(2) $(2/p) = (-1)^{(p^2-1)/8}$.

Proof: Let $r_1, r_2, \ldots, r_{(p-1)/2}$ be integers such that $1 \le r_i \le p-1$, and $r_i \equiv ia \pmod p$. Then

$$\begin{aligned} a &= p[a/p] + r_1 \\ 2a &= p[2a/p] + r_2 \\ &\vdots \\ \frac{1}{2}(p-1)a &= p\left[\frac{1}{2}(p-1)a/p\right] + r_{(p-1)/2}. \end{aligned}$$

Now $1 + 2 + 3 + \cdots + (p-1)/2 = (p^2 - 1)/8$ (recall, or prove, that $1+2+3+\cdots+n = n(n+1)/2$). So, adding up these equations we get

$$\begin{aligned} \frac{1}{8}(p^2 - 1)a &= pM + r_1 + \cdots + r_{(p-1)/2} \\ &= pM + a_1 + \cdots + a_{(p-1)/2} \qquad (2) \end{aligned}$$

in the notation of the proof of Gauss's lemma. On the other hand, $p-a_1, \ldots, p-a_k, a_{k+1}, \ldots, a_{(p-1)/2}$ is a rearrangement of $1, 2, \ldots, (p-1)/2$, so

$$\frac{1}{8}(p^2-1) = 1+2+\cdots+(p-1)/2$$
$$= kp - a_1 - \cdots - a_k + a_{k+1} + \ldots + a_{(p-1)/2}.$$

Subtracting this from (2), we get

$$\frac{1}{8}(p^2-1)(a-1) = p(M-k) + 2a_1 + \cdots + 2a_k$$
$$\equiv M - k \pmod 2.$$

If a is odd, we have $M \equiv k \pmod 2$, and so $(-1)^M = (-1)^k = (a/p)$. If $a = 2$, then $M = 0$ (why?), and we get $(p^2-1)/8 \equiv -k \equiv k \pmod 2$, so $(2/p) = (-1)^{(p^2-1)/8}$.

6.2 PROBLEMS

1. Verify Gauss's lemma for $p = 7$, $a = 3$; $p = 11$, $a = 3$; $p = 17$, $a = 15$; $p = 19$, $a = 2$.
2. Verify the formula $(2/p) = (-1)^{(p^2-1)/8}$ for $p = 3, 5, 7, 11, 13, 17$, and 19.
3. Verify the formula $(a/p) = (-1)^M$ for $p = 7$, $a = 3$; $p = 7$, $a = 10$; $p = 11$, $a = 3$; $p = 11$, $a = -9$.
4. Show that if p is an odd positive prime and $a = p + 2$, then

$$[a/p] + [2a/p] + \cdots + \left[\frac{1}{2}(p-1)a/p\right] = (p^2-1)/8.$$

   Use this to deduce part 2 of the corollary from part 1.
5. For what primes is -2 a quadratic residue?
6. Show that if p is an odd prime, then $(2/p) = 1$ if and only if $p = \pm 1 \pmod 8$.

6.3 THE RECIPROCITY LAW

How do we proceed to evaluate the Legendre symbol (n/p) for p a prime? First of all, we may replace n by the positive number n_o less than

p that is congruent to n modulo p. Theorem 36 (part 1) then tells us that (n_o/p) is a product of terms (q/p) where q is a prime factor of n_o. But how do we go about evaluating a term like (19/113)? Certainly not by calculating 19^{56} (mod 113) à la part 2 of Theorem 36, nor by looking at the first 56 multiples of 19 (mod 113) as in Gauss's lemma. Neither of these important facts provides a very useful computational tool. The pearl of the theory of quadratic residues relates (19/113) to (113/19) in an elegant and useful way.

THEOREM 37. (Reciprocity) If p and q are distinct odd positive primes, then

$$(p/q)(q/p) = (-1)^e$$

where $e = \frac{1}{2}(p-1) \cdot \frac{1}{2}(q-1)$

Proof: Let $M = [q/p] + [2q/p] + \cdots + \left[\frac{1}{2}(p-1)q/p\right]$

$$N = [p/q] + [2p/q] + \cdots + \left[\frac{1}{2}(q-1)p/q\right].$$

Then, by the corollary to Gauss's lemma, $(p/q)(q/p) = (-1)^M(-1)^N = (-1)^{M+N}$, so it will suffice to show that $M + N = \frac{1}{2}(p-1) \cdot \frac{1}{2}(q-1)$. Consider the rectangle R in the plane with vertices (0, 0), (p/2, 0), (0, q/2), (p/2, q/2).

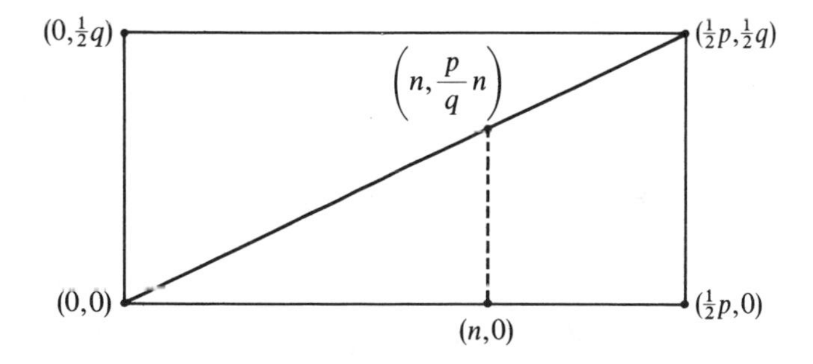

Call a point (x, y) a *lattice point* if x and y are integers. Then R encloses $\frac{1}{2}(p-1)\cdot\frac{1}{2}(q-1)$ lattice points (not counting the ones on the perimeter). The diagonal from $(0, 0)$ to $(p/2, q/2)$ has the equation $y = xq/p$ and so, for any positive integer n, $[np/q]$ is the number of lattice points on the vertical line through $(n, 0)$ that lie on or below the diagonal. But there are no lattice points *on* the diagonal since if $m = nq/p$, then $q/p = m/n$ where $n < p/2$; but q/p is in lowest terms. Thus, M is the number of lattice points below the diagonal. Similarly, N is the number of lattice points above the diagonal, and hence $N + M = \frac{1}{2}(p-1)\cdot\frac{1}{2}(q-1)$.

The reciprocity law may look a little more forbidding than it actually is. The right-hand side is 1 if and only if the exponent is even. This occurs precisely when either $(p-1)/2$ or $(q-1)/2$ is even. But to say that $(p-1)/2$ is even is to say that $p-1$ is divisible by 4, or $p \equiv 1 \pmod 4$. Hence, $(p/q)(q/p) = 1$, that is $(p/q) = (q/p)$, *unless* $p \equiv q \equiv 3 \pmod 4$, in which case $(p/q) = -(q/p)$.

Let's see how we can use the reciprocity law to evaluate $(19/113)$. Since $113 \equiv 1 \pmod 4$, we have $(19/113) = (113/19)$ by the reciprocity law. Thus, $(19/113) = (113/19) = (18/19) = (-1/19) = -1$ since $19 \equiv 3 \pmod 4$.

6.3 PROBLEMS

1. Verify the reciprocity law for $p = 3$, $q = 5$; $p = 3$, $q = 7$; $p = 5$, $q = 29$; $p = 7$, $q = 29$.
2. Evaluate: $(17/113)$; $(37/97)$; $(29/43)$; $(19/47)$.
3. For what primes is 5 a quadratic residue? Find four such primes and check.
4. For what primes is 3 a quadratic residue? Find four such primes and check.

6.4 THE JACOBI SYMBOL

It is convenient to extend the domain of the Legendre symbol (a/p) to allow p to be any odd number (not necessarily prime). If P is an odd positive integer and a is an integer relatively prime to P, we define the *Jacobi symbol* (a/P) by

$$(a/1) = 1$$
$$(a/P) = (a/p_1)(a/p_2)\cdots(a/p_s)$$

for $P = p_1 p_2 \ldots p_s$, p_i prime. Note that the Jacobi symbol agrees with the Legendre symbol when P is a prime, justifying the use of the same notation. Observe also that if a is a quadratic residue of P then a is also a quadratic residue of $p_1, p_2, \ldots, p_s$ (why?), so $(a/P) = 1$. The converse fails to hold since a may be a nonresidue of an even number of the p_i, and hence a nonresidue of P, and still have $(a/P) = 1$. Nevertheless, the Jacobi symbol is an important theoretical and computational tool.

THEOREM 38. Let a and b be integers that are relatively prime to the positive odd integers P and Q. Then

(1) $(a/P)(a/Q) = (a/PQ)$
(2) $(a/P)(b/P) = (ab/P)$
(3) If $a \equiv b \pmod{P}$, then $(a/P) = (b/P)$.

Proof: Immediate consequence of the definition of the Jacobi symbol and the properties of the Legendre symbol.

The reciprocity law extends to the Jacobi symbol.

THEOREM 39. If P and Q are positive relatively prime odd integers, then

(1) $(-1/P) = (-1)^{(P-1)/2}$
(2) $(2/P) = (-1)^{(P^2-1)/8}$
(3) $(P/Q)(Q/P) = (-1)^e$.

where $e = \frac{1}{2}(P-1)\cdot\frac{1}{2}(Q-1)$.

Proof: Let $P = p_1 p_2 \cdots p_s$ and $Q = q_1 q_2 \cdots q_t$ where the p_i and q_j are prime. Then

$$\begin{aligned}(-1/P) &= (-1/p_1)(-1/p_2) \cdots (-1/p_s) \\ &= (-1)^{(p_1-1)/2}\,(-1)^{(p_1-1)/2} \ldots (-1)^{(p_s-1)/2} = (-1)^k\end{aligned}$$

where $k = \Sigma(p_i - 1)/2$. But

$$\begin{aligned}P &= (1 + (p_1 - 1))\,(1 + (p_2 - 1)) \cdots (1 + (p_s - 1)) \\ &\equiv 1 + 2k \pmod 4\end{aligned}$$

since $p_i - 1$ is even. Hence, $k \equiv (P-1)/2 \pmod 2$, so $(-1)^k = (-1)^{(p-1)/2}$. Similarly,

$$\begin{aligned}(2/P) &= (2/p_1)\,(2/p_2) \cdots (2/p_s) \\ &= (-1)^{(p_1^2-1)/8}(-1)^{(p_2^2-1)/8} \cdots (-1)^{(p_s^2-1)/8} = (-1)^h\end{aligned}$$

where $h = \Sigma\,(p_i^2 - 1)/8$. But

$$\begin{aligned}P^2 &= (1 + (p_1^2 - 1))(1 + (p_2^2 - 1)) \cdots (1 + (p_s^2 - 1)) \\ &\equiv 1 + 8h \pmod{64}\end{aligned}$$

since $p_i^2 - 1$ is divisible by 8. Hence, $h \equiv (P^2 - 1)/8 \pmod 8$, so $(-1)^h = (-1)^{(P^2-1)/8}$. Finally,

$$(P/Q) = (P/q_1)\,(P/q_2) \cdots (P/q_t) = \prod (p_i/q_j)$$

where the product is over $i = 1, 2, \ldots, s, j = 1, 2, \ldots, t$. Similarly, $(Q/P) = \prod (q_j/p_i)$. Thus,

$$(P/Q)(Q/P) = \prod (p_i/q_j)(q_j/p_i) = (-1)^n$$

where

$$\begin{aligned}n = \sum_{i,j} \frac{1}{2}(p_i - 1) \cdot \frac{1}{2}(q_j - 1) &= \sum_i \frac{1}{2}(p_i - 1) \sum_j \frac{1}{2}(q_j - 1) \\ &\equiv \frac{1}{2}(P-1) \cdot \frac{1}{2}(Q-1) \pmod 2.\end{aligned}$$

One big computational advantage to the Jacobi symbol is that it allows evaluation of the Legendre symbol without factoring anything except 2's. This is a big help when the factorization is not obvious, and often makes things simpler even when it is. Compare

$$(35/79) = (5/79)(7/79) = -(79/5)(79/7) = -(4/5)(2/7) = -(2/7) = -1$$

to

$$(35/79) = -(79/35) = -(9/35) = -1.$$

Or consider

$$\begin{aligned}(667/919) &= -(919/667) = -(252/667) = -(4/667)(63/667) \\ &= (667/63) = (37/63) = (63/37) = (26/37) = (2/37)(13/37) \\ &= -(37/13) = -(11/13) = -(13/11) = -(2/11) = 1,\end{aligned}$$

which does not require finding out that $667 = 23 \cdot 29$.

6.4 PROBLEMS

1. Prove Theorem 38.
2. Show that the Jacobi symbol may be interpreted as a homomorphism from U_P to $\{1, -1\}$.
3. Let P and Q be odd positive integers.
   (a) Show that $(a/P) = 1$ if a is a square and $(a, P) = 1$.
   (b) Show that $(-1/P) = 1$ if and only if $P \equiv 1 \pmod 4$.
   (c) Show that $(2/P) = 1$ if and only if $P \equiv \pm 1 \pmod 8$.
   (d) Show that $(P/Q) = -(Q/P)$ if and only if $P \equiv Q \equiv 3 \pmod 4$.
4. Evaluate the following with and without the Jacobi symbol: $(21/29)$, $(77/97)$, $(221/257)$.

CHAPTER VII

ALGEBRAIC NUMBER FIELDS

7.1 ALGEBRAIC NUMBERS

In our investigation of when an integer is the sum of two squares, we introduced Gaussian numbers to get an insight into the problem and to aid in its solution. Although we were only interested in integers, the proper domain of number theory, we were led to examine other kinds of numbers. This is a natural, and even inevitable, phenomenon. For example, if we probe the integers hard enough, we will talk about rational numbers now and then even though we could avoid them by sacrificing some clarity and elegance. The numbers we will be drawn to examine will be those that have some arithmetical relation to the integers.

DEFINITION. A complex number α is called an *algebraic number* if there is some nonzero polynomial f with rational coefficients such that $f(\alpha) = 0$. More generally, if F is any field of complex numbers, we say that α is *algebraic over F* if there is a nonzero polynomial f with coefficients in F such that $f(\alpha) = 0$.

Why this definition? Let's examine what we might mean by "the number α is arithmetically related to the integers." If we perform the ordinary arithmetic operations of addition, subtraction, multiplication, and division on the integers, we end up with the rational numbers, no more and no less. The rational numbers are certainly intimately related to the integers. However, we may be able to perform arithmetic operations on an irrational number and end up with a rational number. For example, $\sqrt{2}$ is not rational, but $\sqrt{2} \cdot \sqrt{2} = 2$ is an integer. It is clear that $\sqrt{2}$ *is* related arithmetically to the integers.

In this manner, we are led to consider numbers α that satisfy some equation involving only integers and the arithmetic operations: $\sqrt{2}$ satisfies the equation $x^2 = 2$. Any such equation can be put in the form $G(x) = 0$ where $G(x)$ is an expression involving only the integers, the variable x, and the arithmetic operations. But any such expression $G(x)$ can be reduced to the form $G(x) = f(x)/g(x)$ where f and g are polynomials with rational coefficients. If $G(\alpha) = 0$, then $f(\alpha) = 0$; hence the definition. The requirement that f be nonzero gets rid of trivialities like $\pi/\pi = 1$, which certainly doesn't say anything about how π is related to the integers.

The Gaussian numbers are all algebraic numbers. Indeed, if $\alpha = a + bi$ where a and b are rational numbers, then $\alpha^2 - 2a\alpha + (a^2 + b^2) = 0$, so $f(\alpha) = 0$ where $f(x) = x^2 - 2ax + (a^2 + b^2)$ is a nonzero polynomial with rational coefficients. All square roots, cube roots, etc., of rational numbers are algebraic numbers: $\sqrt{a}$ satisfies the equation $x^2 - a = 0$, $\sqrt[3]{a}$ satisfies the equation $x^3 - a = 0$, and so on.

If α is algebraic over a field F there are many polynomials f with coefficients in F such that $f(\alpha) = 0$; for example, $\sqrt{2}$ is a root of the polynomial $x^3 + x^2 - 2x - 2$ as well as $x^2 - 2$. The most "efficient" polynomials are those of smallest degree, and among these there is a unique one whose leading coefficient is 1. We can find one with leading coefficient 1 simply by dividing any polynomial that has α as a root by its leading coefficient. On the other hand, if f and g are polynomials of smallest degree with leading coefficient 1 such that $f(\alpha) = g(\alpha) = 0$, then $f(x) - g(x)$ is a polynomial of smaller degree satisfied by α and hence, by the minimality of the degrees of f and g, must be zero, so $f(x) = g(x)$. This unique polynomial is called the *minimal polynomial* of α over F. We say that α is of *degree n* over F if its minimal polynomial over F has degree n.

If α is a complex number and F is a field of complex numbers, we denote by $F[\alpha]$ the set of all complex numbers of the form

$$a_k\alpha^k + a_{k-1}\alpha^{k-1} + \cdots + a_1\alpha + a_0$$

where the a_i are in F and k is a nonnegative integer. Thus, $F[\alpha]$ consists of all those numbers which can be "built up" from F and α by addition and multiplication. It is not hard to show that $F[\alpha]$ is an integral domain; we leave that as an exercise.

THEOREM 40. If α is of degree n over F, then every element of $F[\alpha]$ may be written uniquely as $a_0 + a_1\alpha + \cdots + a_{n-1}\alpha^{n-1}$ where the a_i are in F.

Proof: We must show that any element of $F[\alpha]$ can be so represented and that the representation is unique. Let $f(x) = x^n + c_{n-1}x^{n-1} + \cdots + c_1x + c_0$ be the minimal polynomial of α over F. Since $f(\alpha) = 0$, we can write $\alpha^n = -c_{n-1}\alpha^{n-1} - \cdots - c_1\alpha - c_0$. This is the key; it allows us to write α^n in terms of smaller powers of α. Let β be any element of $F[\alpha]$. Then $\beta = a_k\alpha^k + a_{k-1}\alpha^{k-1} + \cdots + a_1\alpha + a_0$ for some nonnegative integer k and elements $a_i \in F$. Among all such ways of writing β, choose one which minimizes k. If $k < n$, we are through. Suppose $k \geq n$. Then $\alpha^k = \alpha^{k-n}\alpha^n = -c_{n-1}\alpha^{k-1} - \cdots - c_0\alpha^{k-n}$. Plugging this in for α^k in the expression for β, we get an expression for β involving only powers of α less than k, contradicting our choice of k.

To prove uniqueness, we capitalize on the minimality of n. Suppose

$$a_0 + a_1\alpha + \cdots + a_{n-1}\alpha^{n-1} = b_0 + b_1\alpha + \cdots + b_{n-1}\alpha^{n-1}.$$

Then

$$(a_0 - b_0) + (a_1 - b_1)\alpha + \cdots + (a_{n-1} - b_{n-1})\alpha^{n-1} = 0.$$

But this says that α is a root of the polynomial

$$f(x) = (a_0 - b_0) + (a_1 - b_1)x + \cdots + (a_{n-1} - b_{n-1})x^{n-1},$$

which has degree less than n, unless all the coefficients are 0. Hence, $f = 0$; that is, $a_0 = b_0$, $a_1 = b_1$, $\ldots$, $a_{n-1} = b_{n-1}$.

7.1 PROBLEMS

1. Show that every algebraic number is the root of some nonzero polynomial with integer coefficients.
2. Show that a Gaussian number $\alpha = a + bi$ is algebraic of degree 2 if and only if $b \neq 0$.
3. Show that the minimal polynomial of $\alpha = \sqrt[3]{2}$ is $x^3 - 2$ (show that α is irrational; then fiddle with the equations $\alpha^3 = 2$ and $\alpha^2 + a\alpha + b = 0$).
4. Show that if F is a field of complex numbers and α is a complex number then $F[\alpha]$ is an integral domain (use the fact that the complex numbers are a field).
5. Show that if F is a field of complex numbers and α is a complex number, then any ring of complex numbers containing F and α contains $F[\alpha]$—that is, $F[\alpha]$ is the smallest ring containing F and α.
6. Show that $F = Q[\sqrt{3}]$ is a field (Q is the field of rational numbers—"Q" for *Q*uotients of integers). What is the degree of i over F?
7. Let α be an algebraic number. Show that the set of all polynomials f with rational coefficients such that $f(\alpha) = 0$ is an ideal in the ring of all such polynomials.
8. Let $\alpha \neq 0$ be algebraic over the field F. Show that $1/\alpha$ is algebraic over F of the same degree as α. (If $a_n\alpha^n + a_{n-1}\alpha^{n-1} + \cdots + a_1\alpha + a_0 = 0$, then $a_n + a_{n-1}(1/\alpha) + \cdots + a_1(1/\alpha)^{n-1} + a_0(1/\alpha)^n = 0$.)

7.2 MORE ABOUT POLYNOMIALS

Since we shall be dealing with roots of polynomials, it behooves us to develop the theory of polynomials a little more. If F is a field, we denote the ring of polynomials in the variable x with coefficients in F by $F[x]$, which is consistent with our previous notation $F[\alpha]$, where α was a number. Suppose α is algebraic over F. Consider the set $I = \{g \in F[x] \mid g(\alpha) = 0\}$. Clearly I is an ideal in $F[x]$ (check it out). We know that every ideal in Z is principal; is every ideal in $F[x]$ principal? It is, and the reason is the same: the division algorithm.

THEOREM 41. If F is a field and I is an ideal in the ring $F[x]$, then I is principal, and any nonzero polynomial in I of smallest degree is a generator.

Proof: If $I = \{0\}$, then I is principal and 0 is a generator. If $I \neq \{0\}$, let f be a nonzero polynomial in I of smallest degree. Then every multiple of f is in I; we must show that, conversely, every element of I is a multiple of f. Let $g \in I$. Then, by Theorem 24, the division algorithm for polynomials, there exist polynomials $q, r \in F[x]$ such that $g = qf + r$, and $r = 0$ or the degree of r is less than the degree of f. Since g and f are in I, so is $r = g - qf$. Hence, by the minimality of the degree of f, we must have $r = 0$, and therefore $g = qf$ is a multiple of f. Compare this to the proof of Theorem 3.

COROLLARY 1. Let α be algebraic over F and f the minimal polynomial of α over F. If $g \in F[x]$ then $g(\alpha) = 0$ if and only if g is a multiple of f.

Proof: Let I be the set of all polynomials in $F[x]$ that have α as a root. Then I is a nonzero ideal in $F[x]$. By definition, the minimal polynomial f of α is a nonzero polynomial in I of smallest degree. Hence, by Theorem 41, $g \in I$ if and only if g is a multiple of f.

This is more than we knew before; not only is the minimal polynomial f of α over F an element of $F[x]$ of smallest degree that has α as a root, but any element of $F[x]$ that has α as a root is a *multiple* of f. Compare this to the fact that the greatest common divisor of two integers is not only bigger than any other common divisor, but in fact is a multiple of any common divisor.

It is easy to verify that the units of the ring $F[x]$ are precisely the nonzero constant polynomials—that is, the elements of degree zero. If $f(x)$ is a prime in the ring $F[x]$, that is, if f cannot be written as $f = gh$ unless g or h is a constant, then f is said to be *irreducible* over F. It is essential to specify the field F when talking about irreducible polynomials; for example, $x^2 + 2$ is irreducible over Q (see Problem 3) but not irreducible over the complex numbers, where

$$x^2 + 2 = (x + i\sqrt{2})(x - i\sqrt{2}).$$

The minimal polynomial f of α over F completely describes the interaction between α and F. Any root of f behaves exactly like α does.

COROLLARY 2. Let α be algebraic over F and f the minimal polynomial of α over F. Then f is irreducible. Moreover, if $f(\beta) = 0$, then f is the minimal polynomial of β over F, and if $g \in F[x]$, then $g(\alpha) = 0$ if and only if $g(\beta) = 0$.

Proof: If $f(x) = a(x)b(x)$, then $0 = f(\alpha) = a(\alpha)b(\alpha)$, so either $a(\alpha) = 0$ or $b(\alpha) = 0$. But, since f is the *minimal* polynomial of α, this implies that either b or a has degree zero. If $f(\beta) = 0$, then, by Corollary 1, f is a multiple of the minimal polynomial of β. But, since f is irreducible, f must *be* the minimal polynomial of β. Finally, since α and β have the same minimal polynomial, and $g(\theta) = 0$ if and only if g is a multiple of the minimal polynomial of θ, then $g(\alpha) = 0$ if and only if $g(\beta) = 0$.

We have seen (Corollary 2 to Theorem 24) that a polynomial of degree n has at most n roots. In a certain sense, every polynomial has precisely n roots, if we count some of the roots more than once. For example, the polynomial $x^2 - 2x + 1$ has only the number 1 as a root, but $x^2 - 2x + 1 = (x - 1)^2$ and, if we agree to count the root 1 twice (which is not totally unreasonable since, in some sense, $(x - 1)^2$ is zero "twice as hard" at 1 than, say, $(x - 1)(x - 7)$ is), then we can say that $x^2 - 2x + 1$ has two roots, which both happen to be 1; we say that 1 is a *multiple root* of $x^2 - 2x + 1$, and that its *multiplicity* is 2. This may appear a little shady, but it is convenient.

Perhaps this can be made a little more plausible by recalling Corollary 1 to Theorem 24: if F is a field, $f(x) \in F[x]$ and $\alpha \in F$, then α is a root of f if and only if $(x - \alpha)$ divides $f(x)$. If we think of this as a *definition* of what it means for α to be a root of f, then the following definition is reasonable.

DEFINITION. If F is a field, $f(x) \in F[x]$, and $\alpha \in F$, then α is a *root of multiplicity m* of f if $(x - \alpha)^m$ divides $f(x)$ but $(x - \alpha)^{m+1}$ does not divide $f(x)$.

Does a polynomial have to have any roots at all? Some of them don't, if we require the roots to lie in a specified field. For example, neither $x^2 - 2$ nor $x^2 + 1$ have roots in Q. However, by looking in a large enough field, such as the real numbers for $x^2 - 2$ or the Gaussian numbers for $x^2 + 1$, we can find a root. A basic theorem in algebra says that if F is a field and $f(x) \in F[x]$ has degree greater than zero, then there is a field K containing F in which f has a root. This is a rather abstract theorem and a little hard to appreciate. More down to earth, but harder to prove, is the following theorem.

THEOREM (The fundamental theorem of algebra). If $\mathscr{C}$ is the field of complex numbers and $f \in \mathscr{C}[x]$ has degree greater than zero, then f has a root in $\mathscr{C}$.

We shall not prove this theorem here; the concepts and techniques involved are totally different from those we have been dealing with. Indeed, despite its name, it is not a theorem of algebra at all! We quote it here merely to indicate that there is no point looking for algebraic numbers anywhere other than among the complex numbers. This is a consequence of the following corollary.

COROLLARY. If f is a nonzero polynomial with complex coefficients, then f can be written, uniquely except for order of the factors, as $f(x) = a(x - r_1)(x - r_2)\cdots(x - r_n)$ where a, r_1, r_2, $\ldots$, r_n are complex numbers.

Proof: Suppose not. Let f be a polynomial of smallest degree for which the theorem fails. Then, by the fundamental theorem of algebra, f has a root r in the complex numbers. Hence, by Corollary 1 to Theorem 24, $f(x) = (x - r)g(x)$ for some polynomial g with complex coefficients, and since $\mathscr{C}[x]$ is an integral domain, the polynomial g is unique. Since the degree of g is less than the degree of f, we can write g uniquely in the desired form, and hence we can write $f(x) = (x - r)g(x)$ in the desired form. Since r is a root of f, *any* such expression for f must include a factor $x - r$ (why?). The remaining factors must multiply out to g, since g is unique. But there is only one such way to factor g, by the minimality of the degree of f, so the factorization of f is unique.

In particular, if f is a polynomial with rational coefficients and $f(\alpha) = 0$ where α is an element of some field containing the complex numbers, then $f(\alpha) = a(\alpha - r_1)(\alpha - r_2)\cdots(\alpha - r_n) = 0$ and so $\alpha = r_i$, a complex number, for some i.

One of the nice things about irreducible polynomials, at least those with complex coefficients, is that they cannot have multiple roots. We shall have occasion to use this fact.

THEOREM 42. Let F be a field of complex numbers. If $f \in F[x]$ is irreducible over F, then f has no multiple roots.

Proof: We may as well assume that the leading coefficient of f is 1. Suppose f has a multiple root $\alpha \in \mathscr{C}$. Then f is divisible by the minimal polynomial of α over F and so, since f is irreducible over F, it must *be* the minimal polynomial of α over F. Then $f(x) = (x - \alpha)^2 g(x)$ for some $g \in \mathscr{C}[x]$ (*not* $F[x]$). If $f(x) = a_n x^n + a_{n-1}x^{n-1} + \cdots + a_1 x + a_0$, define $f'(x)$ to be $na_n x^{n-1} + (n-1)a_{n-1}x^{n-2} + \cdots + 2a_2 x + a_1$. Then $f'(x) \in F[x]$ and $f'(x) = 2(x - \alpha)g(x) + (x - \alpha)^2 g'(x)$, just like differentiation in calculus, although we do not have to talk about limits (see

Problem 7). Clearly $f'(\alpha) = 0$. But the degree of f' is less than the degree of f, which is the minimal polynomial of α over F, a contradiction.

7.2 PROBLEMS

1. Let F be a field and f a monic irreducible polynomial with coefficients in F. Show that if $f(\alpha) = 0$, then f is the minimal polynomial of α over F.
2. Show that any polynomial of degree 1 with coefficients in a field F is irreducible over F.
3. Show that a polynomial of degree 2 or 3 with coefficients in a field F is irreducible over F if and only if it has no roots in F.
4. Use Problem 3 to find irreducible polynomials of degree 2 and 3 with coefficients in Z_2, Z_3, and Z_5.
5. Find an example of a polynomial of degree 4 with rational coefficients which is not irreducible over Q yet has no rational roots.
6. (Fundamental theorem of arithmetic for $F[x]$). Let F be a field. We wish to prove things about $F[x]$ that we have already proved about Z. To translate, keep in mind that *irreducible polynomials* correspond to *prime numbers, nonzero* "*constant polynomials*" (polynomials of degree zero) correspond to the numbers 1 and -1 (the units), and *degree* corresponds to *absolute value*. Note also that *monic* corresponds to *positive*, singling out a representative for a class of associates.
   (a) State and prove the analogue of Theorem 2 for $F[x]$.
   (b) Let f and g be elements of $F[x]$. By a *greatest common divisor* of f and g we mean a monic polynomial of greatest degree which divides both f and g. State and prove the analogue of Theorem 4 for $F[x]$. Conclude that the greatest common divisor is unique.
   (c) State and prove the analogue of Theorem 5 for $F[x]$.
   (d) State and prove the fundamental theorem of arithmetic for $F[x]$. (The statement is clearer if you restrict yourself to monic polynomials; otherwise, see Problem 12 of section 5.2.)

7. (Formal differentiation) Let F be a field and $f \in F[x]$. If $f = a_n x^n + a_{n-1}x^{n-1} + \cdots + a_1 x + a_0$, *define* f' to be the polynomial $na_n x^{n-1} + (n-1)a_{n-1}x^{n-2} + \cdots + 2a_2 x + a_1$. Show that if $f, g \in F[x]$ and $r \in F$, then
   (a) $(rf)' = rf'$
   (b) $(f+g)' = f' + g'$
   (c) $(fg)' = f'g + fg'$
   (d) $(f^n)' = nf^{n-1}f'$, for n a positive integer (choose the least n for which it fails and use (c)).

7.3 VECTOR SPACES

If α is an algebraic number of degree n, then any element of $Q[\alpha]$ can be written uniquely as $a_0 + a_1\alpha + \cdots + a_{n-1}\alpha^{n-1}$, where $a_i \in Q$. This representation is strongly reminiscent of the representation of elements in a direct sum of groups. In fact, it shows that $Q[\alpha]$, as a group under addition, is a direct sum of the groups $Q, Q\alpha, \ldots, Q\alpha^{n-1}$, where $Q\alpha^i = \{q\alpha^i \mid q \in Q\}$. It is easily seen that $Q\alpha^i$ is isomorphic to Q under the isomorphism $f(q\alpha^i) = q$. Hence, as an abelian group, $Q[\alpha]$ is simply a direct sum of n copies of Q.

Of course, $Q[\alpha]$ also has the structure of a ring, since we can multiply any two elements; but this will not concern us right now. The additional structure we wish to consider now is simply the ability to multiply elements of $Q[\alpha]$ by elements of Q. More generally, if F is any field of complex numbers, then a subgroup V of the complex numbers is said to be a *vector space over* F if for any $a \in F$ and $v \in V$ we have $av \in V$. In this language, we say that $Q[\alpha]$ is a vector space over Q. The all-important characteristic of a vector space is its "size." We measure the size of a vector space by a positive integer called its *dimension*.

DEFINITION. If V is a vector space over the field F, then we say that V has *dimension* n over F, and write $\dim_F V = n$, if there exist elements $v_1, v_2, \ldots, v_n$ in V such that every element of V can be written uniquely in the form $a_1v_1 + a_2v_2 + \cdots + a_nv_n$ where the a_i are in F. The elements $v_1, v_2, \ldots, v_n$ are said to form a *basis* for V over F. If there is no basis for V over F, we say that V is *infinite dimensional* over F and write $\dim_F V = \infty$.

Thus, Theorem 40 says that $Q[\alpha]$ has dimension n over Q and the elements $1, \alpha, \alpha^2, \ldots, \alpha^{n-1}$ are a basis for $Q[\alpha]$ over Q. There are problems involved with this definition. Generally, a given vector space has many different bases, and if the bases didn't all have the same number of elements, then the vector space wouldn't have a unique dimension.

The Gaussian numbers $Q[i]$ are a vector space over Q. One basis is $\{1, i\}$ since every Gaussian number may be written uniquely as $a + bi$ where $a, b \in Q$. But the numbers $1 + i$ and $1 - i$ are also a basis for $Q[i]$. Indeed, any Gaussian number $a + bi$ may be written as $c(1 + i) + d(1 - i)$, where $c = (a + b)/2$ and $d = (a - b)/2$; and this representation is unique, for if $c(1 + i) + d(1 - i) = a + bi$ then $c = (a + b)/2$ and $d = (a - b)/2$. Although the basis $\{1, i\}$ may seem more natural, it is certainly not the only one. The first order of business is to show that any two bases have the same number of elements. To do this, we split the property of being a basis into two parts. We call a sum of the form $a_1v_1 + a_2v_2 + \cdots + a_nv_n$, where the a_i are in F and the v_i are in V, a *linear combination* of the v_i (with coefficients in F).

DEFINITION. If V is a vector space over F, we say that the vectors $v_1, v_2, \ldots, v_n$ *span* V if every element of V can be written (not necessarily uniquely) as a linear combination of the v_i.

DEFINITION. If V is a vector space over F, we say that the vectors $v_1, v_2, \ldots, v_n$ are *linearly independent* if whenever a linear combination $a_1v_1 + a_2v_2 + \cdots + a_nv_n = 0$ then $a_1 = a_2 = \cdots = a_n = 0$.

Notice that for $v_1, v_2, \ldots, v_n$ to be linearly independent means that there is only one way to represent 0 as a linear combination of the v_i. From this, it follows that there is only one way to represent any element of V as a linear combination of the v_i. Indeed, if

$$v = a_1v_1 + a_2v_2 + \cdots + a_nv_n = b_1v_1 + b_2v_2 + \cdots + b_nv_n,$$

then $(a_1 - b_1)v_1 + (a_2 - b_2)v_2 + \cdots + (a_n - b_n)v_n = 0$, so $a_1 = b_1, a_2 = b_2, \ldots, a_n = b_n$. Thus, $v_1, v_2, \ldots, v_n$ is a basis for V if and only if the v_i span V and are linearly independent.

THEOREM 43. Let V be a vector space over the field F. If the vectors $u_1, u_2, \ldots, u_n$ span V and the vectors $v_1, v_2, \ldots, v_m$ are linearly independent, then $m \leq n$.

Proof: Since the u_i span V, we can write $v_1 = a_1 u_1 + a_2 u_2 + \cdots + a_n u_n$. If all the $a_i = 0$, then $v_1 = 0$, which is impossible, for then $1 \cdot v_1 + 0 \cdot v_2 + \ldots + 0 \cdot v_m = 0$, contradicting the independence of the v_i. Hence, some a_i is nonzero, and by relabeling if necessary, we may assume that $a_1 \neq 0$. Thus, $u_1 = (1/a_1)v_1 - (a_2/a_1)u_2 - \cdots - (a_n/a_1)u_n$, and hence $v_1, u_2, \ldots, u_n$ spans V, for we may use this expression for u_1 to write any linear combination of $u_1, u_2, \ldots, u_n$ as a linear combination of $v_1, u_2, \ldots, u_n$. In particular, $v_2 = b_1 v_1 + b_2 u_2 + \cdots + b_n u_n$. If $b_2 = b_3 = \cdots = b_n = 0$, then the v_i would not be linearly independent. Hence, some b_i, $i \geq 2$, is nonzero and, relabeling if necessary, we may assume that $b_2 \neq 0$. Then

$$u_2 = (1/b_2)v_2 - (b_1/b_2)v_1 - (b_3/b_2)u_3 - \cdots - (b_n/b_2)u_n,$$

and so $v_1, v_2, u_3, \ldots, u_n$ span V. Continuing this argument, replacing the u's by v's, we see that if $m > n$, then $v_1, v_2, \ldots, v_n$ would span V, and so v_{n+1} would be a linear combination of $v_1, v_2, \ldots, v_n$, contradicting the independence of the v's.

COROLLARY 1. If V is a vector space, then any two bases of V have the same number of elements.

Proof: If $u_1, u_2, \ldots, u_n$ is a basis and $v_1, v_2, \ldots, v_m$ is a basis, then, by Theorem 43, $m \leq n$, since the u_i span V and the v_i are linearly independent. On the other hand, the v_i span V and the u_i are linearly independent so, by Theorem 43, $n \leq m$. Hence, $n = m$.

COROLLARY 2. If V is an n-dimensional vector space, then no set of $n + 1$ vectors can be linearly independent.

Proof: If $v_1, v_2, \ldots, v_{n+1}$ were linearly independent, then Theorem 43 says that $n + 1 \leq n$, which it is not.

COROLLARY 3. A number α is algebraic over the field F if and only if α is contained in a ring R containing F such that $\dim_F R$ is finite. If $\alpha \in R$, then the degree of α over F is no greater than $\dim_F R$.

Proof: Observe that any ring containing a field is a vector space over that field. If α is algebraic over F, then $\alpha \in F[\alpha]$, which is a finite-dimensional ring over F. If $\alpha \in R$, where $\dim_F R = n$, then the numbers $1, \alpha, \alpha^2, \ldots, \alpha^n$ are $n + 1$ vectors in R and hence are not linearly independent. Therefore, there exist elements $a_i \in F$, not all zero, such that $a_0 + a_1\alpha + \cdots + a_n\alpha^n = 0$—that is, $f(\alpha) = 0$ where $f(x) = a_0 + a_1 x + \cdots + a_n x^n$ is a nonzero polynomial with coefficients in F of degree no greater than n (a_n might be 0).

7.3 PROBLEMS

1. Show that any field F is a one-dimensional vector space over itself.
2. Find two different bases for the vector space $Q[\sqrt{2}]$ over Q.
3. Let V be a vector space over F. Show that the vectors $v_1, v_2, \ldots, v_n$ are linearly independent if and only if none of them can be written as a linear combination of the others.
4. Show that any finite dimensional vector space over a field F is a direct sum (as abelian groups) of one-dimensional vector spaces.
5. Let V be a vector space over F and a an element of F. Show that the vectors $v_1, v_2, \ldots, v_n$ are linearly independent if and only if the vectors $v_1 + av_2, v_2, \ldots, v_n$ are linearly independent. Do the same thing with the words "are linearly independent" replaced by "span V."
6. Let V be an n-dimensional vector space. Show that any n linearly independent vectors form a basis. Show that any n vectors that span V form a basis.
7. Let α be algebraic over the field F. Show that the degree of α over F is $\dim_F F[\alpha]$.

8. Let α and β be algebraic over the field F, of the same degree. Show that if $\beta \in F[\alpha]$, then $F[\alpha] = F[\beta]$.
9. Let $\mathscr{C}$ be the complex numbers and $\mathscr{R}$ be the real numbers. What is the dimension of $\mathscr{C}$ as a vector space over $\mathscr{R}$? Find a basis.

7.4 NUMBER FIELDS

We call $F[\alpha]$ the *ring generated by* α *over* F since it is the smallest ring containing F and α (see Problem 5 of section 7.1). If α is algebraic over F, then $F[\alpha]$ is also a field. Indeed, this property characterizes numbers algebraic over F.

THEOREM 44. A complex number α is algebraic over the field F if and only if $F[\alpha]$ is a field.

Proof: To show that $F[\alpha]$ is a field we must show that for every $\beta \neq 0$ in $F[\alpha]$, there is a $\gamma \in F[\alpha]$ such that $\gamma\beta = 1$. Since $\beta \in F[\alpha]$, β is algebraic over F, by Corollary 3 to Theorem 43. Let

$$f(x) = x^n + a_{n-1}x^{n-1} + \cdots + a_1 x + a_0$$

be the minimal polynomial of β over F. Then $a_0 \neq 0$, for otherwise β would satisfy the polynomial

$$x^{n-1} + a_{n-1}x^{n-2} + \cdots + a_2 x + a_1.$$

Thus,

$$\beta^n + a_{n-1}\beta^{n-1} + \cdots + a_1\beta = -a_0,$$

so

$$\gamma = (-1/a_0)\beta^{n-1} + (-a_{n-1}/a_0)\beta^{n-2} + \cdots + (-a_1/a_0) \in F[\alpha]$$

and $\gamma\beta = 1$.

Conversely, suppose $F[\alpha]$ is a field. Then there is a $\beta \in F[\alpha]$ such that $\beta\alpha = 1$. Let

$$\beta = a_n \alpha^{n-1} + \cdots + a_2 \alpha + a_1$$

where $a_n \neq 0$ and the a_i are in F. Then

$$a_n\alpha^n + \cdots + a_1\alpha + (-1) = 0,$$

so

$$f(x) = a_n x^n + \cdots + a_1 x + (-1)$$

is a nonzero polynomial with coefficients in F such that $f(\alpha) = 0$.

DEFINITION. An *algebraic number field* is a field of complex numbers that is a finite-dimensional vector space over the rational numbers.

Every element α of an algebraic number field F is algebraic, by Corollary 3 to Theorem 43 (it would certainly be a funny definition were this not the case). By Theorem 44, if α is an algebraic number, then $Q[\alpha]$ is an algebraic number field. We shall show in the next section that every algebraic number field is of this form.

Suppose F is an algebraic number field and α is algebraic over F. We have concluded that the elements of F, being algebraic numbers, are fit objects for study; but what about α? Once we agree to study F, the same argument forces us to look at elements like α. More generally, at least at first glance, what about elements that satisfy polynomials whose coefficients are algebraic numbers? And if we look at numbers like these, what about numbers that satisfy polynomials whose coefficients satisfy polynomials whose coefficients satisfy polynomials ...? We may have opened a Pandora's box. Fortunately, the situation is quite simple. The basic setup involves three fields; for example, $F[\alpha]$ is finite dimensional over F and F is finite dimensional over Q.

THEOREM 45. Suppose F, G, and H are three fields such that $F \subseteq G \subseteq H$. If G is finite dimensional over F and H is finite dimensional over G, then H is finite dimensional over F. Moreover, $\dim_F H = \dim_F G \cdot \dim_G H$.

Proof: Let $g_1, g_2, \ldots, g_n$ be a basis for G over F and $h_1, h_2, \ldots, h_m$ be a basis for H over G. We will show that the set of products $g_i h_j$, $i = 1, \ldots, n$, $j = 1, \ldots, m$, is a basis for H over F. First we show that they span H. Suppose $\alpha \in H$. Then $\alpha = \sum_{j=1}^{m} a_j h_j$, where $a_j \in G$, since the h's span H over G. But, since $a_j \in G$, $a_j = \sum_{i=1}^{n} f_{ji} g_i$, where $f_{ji} \in F$, since the g's span G over F. Hence $\alpha = \sum_{j=1}^{m} \sum_{i=1}^{n} f_{ji} g_i h_j$ where $f_{ji} \in F$, so the elements $g_i h_j$ span H over F. To show independence, suppose $\sum_{j=1}^{m} \sum_{i=1}^{n} f_{ji} g_i h_j = 0$. Then, since the h's are linearly independent over G and $\sum_{i=1}^{n} f_{ji} g_i \in G$, we must have $\sum_{i=1}^{n} f_{ji} g_i = 0$ for every j. But $f_{ji} \in F$ and the g's are linearly independent over F, so $f_{ji} = 0$ for every i and j. Hence, the $g_i h_j$ are linearly independent over F.

COROLLARY 1. Let $F \subseteq K$ be fields such that $\dim_F K = n$, and let α be algebraic of degree m over K. Then α is algebraic over F and the degree of α divides $n \cdot m$.

Proof: Theorem 45 applied to the fields $F \subseteq K \subseteq K[\alpha]$ shows that $\dim_F K[\alpha] = n \cdot m$. Hence α is algebraic over F by Corollary 3 to Theorem 43. Now Theorem 45 applied to the fields $F \subseteq F[\alpha] \subseteq K[\alpha]$ shows that $\dim_F F[\alpha]$, which by Theorem 40 is the degree of α, divides $\dim_F K[\alpha] = n \cdot m$.

That settles the first question: we get nothing new if we look at numbers that are algebraic over algebraic number fields. What about numbers that satisfy polynomials whose coefficients are algebraic numbers? Suppose $a_0 + a_1\alpha + \cdots + a_n\alpha^n = 0$, where the a_i are algebraic numbers. If we could put all the a_i inside one algebraic number field, we could use Corollary 1 to conclude that α was algebraic. We extend our notation to allow us to write down the "best try" at putting $a_0, a_1, \ldots, a_n$ in an algebraic number field.

If a_0 and a_1 are complex numbers and F is a field, we denote by $F[a_0, a_1]$ the set of all numbers of the form $\sum f_{ij} a_0{}^i a_1{}^j$ where $f_{ij} \in F$ and i and j are nonnegative integers—that is, $F[a_0, a_1]$ consists of all "polynomials" in a_0 and a_1 with coefficients in F. It is readily seen that $F[a_0, a_1] = (F[a_0])\,[a_1] = (F[a_1])\,[a_0]$. Similarly, we denote by $F[a_0, \ldots, a_n]$ the set of all "polynomials" in $a_0, \ldots, a_n$ with coefficients in F. We say that $F[a_0, \ldots, a_n]$ is the ring generated by $a_0, \ldots, a_n$ over F, for the standard reason (see Problem 1).

COROLLARY 2. If $a_0, \ldots, a_n$ are algebraic over F, then $F[a_0, \ldots, a_n]$ is finite dimensional as a vector space over F.

Proof: Since a_0 is algebraic over F, $F[a_0]$ is finite dimensional over F by Theorem 40. Since a_1 is algebraic over F, it is also algebraic over the field $F[a_0]$; indeed, any polynomial with coefficients in F can also be regarded as a polynomial with coefficients in $F[a_0]$. Hence, $F[a_0, a_1]$ is finite dimensional over $F[a_0]$ and so, by Theorem 45, $F[a_0, a_1]$ is finite dimensional over F. Continuing, a_2 is algebraic over $F[a_0, a_1]$, so $F[a_0, a_1, a_2]$ is finite dimensional over $F[a_0, a_1]$, which is finite dimensional over F so, by Theorem 45, $F[a_0, a_1, a_2]$ is finite dimensional over F, etc.

COROLLARY 3. If a and b are algebraic over F, then so are $a+b$, $a-b$, ab, and, if $b \neq 0$, a/b.

Proof: Since $F[a, b]$ is a field containing a and b, it contains $a+b$, $a-b$, ab, and a/b. Since it is finite dimensional over F, any element in it is algebraic over F by Corollary 3 to Theorem 43.

COROLLARY 4. If α satisfies a nontrivial polynomial whose coefficients are algebraic over F, then α is algebraic over F.

Proof: Suppose $a_0 + a_1\alpha + \cdots + a_n\alpha^n = 0$ where the a_i are algebraic over F and not all zero. Then α is algebraic over the field $F[a_0, \ldots, a_n]$, so $F[a_0, \ldots, a_n, \alpha]$ is finite dimensional over $F[a_0, \ldots, a_n]$, which is finite dimensional over F. Thus, α is an element of the field $F[a_0, \ldots, a_n, \alpha]$ which is finite dimensional over F, by Theorem 45, so α is algebraic over F by Corollary 3 to Theorem 43.

That is a marvelous state of affairs. If we apply arithmetic operations to algebraic numbers we end up with algebraic numbers. Also, any number that satisfies a nontrivial polynomial whose coefficients are algebraic numbers is an algebraic number. Thus, algebraic numbers are the only numbers we will expect to run into in our study of the integers.

7.4 PROBLEMS

1. Let $a_0, \ldots, a_n$ be complex numbers and F a field of complex numbers. Show that $F[a_0, \ldots, a_n]$ is an integral domain containing F and $a_0, \ldots, a_n$, and that if R is any ring containing F and $a_0, \ldots, a_n$, then $F[a_0, \ldots, a_n] \subseteq R$.
2. Show that $Q[i] \subseteq Q[\sqrt{2} + i]$ and $Q[\sqrt{2}] \subseteq Q[\sqrt{2} + i]$.
3. Find a basis for $Q[\sqrt{2} + i]$ over $Q[\sqrt{2}]$; over $Q[i]$; over Q.
4. Show that $Q[\sqrt{2}] \subseteq Q[\sqrt{2} + \sqrt{3}]$ and $Q[\sqrt{3}] \subseteq Q[\sqrt{2} + \sqrt{3}]$.
5. Find the minimal polynomial of $\sqrt{2} + \sqrt{3}$ over Q; over $Q[\sqrt{2}]$; over $Q[\sqrt{3}]$.
6. Show that if α is an algebraic number of degree n and β is an algebraic number of degree m, then $\alpha + \beta$ and $\alpha\beta$ are algebraic numbers of degree at most $n \cdot m$.
7. What is the degree of $i + \sqrt{2}$? $i\sqrt{2}$?
8. Show that if F is an algebraic number field of dimension p over Q, where p is a prime number, then $F = Q[\alpha]$ for any $\alpha \in F$ that is not in Q.

7.5 SIMPLE EXTENSIONS

We raised the question in the last section about when we can write an algebraic number field in the form $Q[\alpha]$ for some algebraic number α. If F is a field and α is algebraic over F, we call the field $F[\alpha]$ a *simple algebraic extension* of F.

THEOREM 46. If F is a field of complex numbers and α and β are algebraic over F, then there is a complex number θ such that $F[\alpha, \beta] = F[\theta]$.

Proof: Let f and g be the minimal polynomials of α and β over F. Let $\alpha = \alpha_1, \alpha_2, \ldots, \alpha_n$ and $\beta = \beta_1, \beta_2, \ldots, \beta_m$ be the roots of f

and g, respectively. Choose a rational number t different from the numbers $(\alpha_i - \alpha)/(\beta_j - \beta)$ for $i = 1, 2, \ldots, n; j = 2, 3, \ldots, m$. Set $\theta = \alpha + t\beta \in F[\alpha, \beta]$. Let $h(x) = f(\theta - tx)$. Then h is a polynomial with coefficients in $F[\theta]$ such that $h(\beta) = 0$ but $h(\beta_j) \neq 0$ for $j = 2, 3, \ldots, m$. If $k(x)$ is the minimal polynomial of β *over* $F[\theta]$ then $k(x)$ divides both $g(x)$ and $h(x)$. Since h and g have only the root β in common, and since g has no multiple roots (it is irreducible over F), then $k(x)$ must be of degree 1, and hence $k(x) = x - \beta$. But $k(x)$ is a polynomial with coefficients in $F[\theta]$, so $\beta \in F[\theta]$. Since $\alpha = \theta - t\beta$, we also have $\alpha \in F[\theta]$, so $F[\alpha, \beta] \subseteq F[\theta]$. Since $\theta \in F[\alpha, \beta]$, $F[\alpha, \beta] = F[\theta]$.

COROLLARY 1. If K is a field of complex numbers that is finite-dimensional over a field $F \subseteq K$, then $K = F[\theta]$ for some θ which is algebraic over F.

Proof: Let $\alpha_1, \ldots, \alpha_n$ be a basis for K over F. Then, by Theorem 46, there is a θ_1 such that $F[\alpha_1, \alpha_2] = F[\theta_1]$. Similarly, there is a θ_2 such that $F[\alpha_1, \alpha_2, \alpha_3] = F[\theta_1, \alpha_3] = F[\theta_2]$. Continuing, we get a θ such that $K = F[\alpha_1, \ldots, \alpha_n] = F[\theta]$. Everything in sight is algebraic over F by Corollary 3 to Theorem 43.

COROLLARY 2. If K is an algebraic number field, then $K = Q[\theta]$ for some algebraic number θ.

7.5 PROBLEMS

1. Show that if $a \neq b$ are rational numbers, then $Q[\sqrt{a}, \sqrt{b}] = Q[\sqrt{a} + \sqrt{b}]$.
2. Let $\alpha = \sqrt{1 + \sqrt{2}}$, $\beta = \sqrt{1 - \sqrt{2}}$.
   (a) Show that $Q[\sqrt{2}] \subseteq Q[\beta]$ and $Q[\sqrt{2}] \neq Q[\beta]$ (β is not real). Conclude that the degree of β is 4.
   (b) Find the minimal polynomial of β and show that it is also the minimal polynomial of α. Hence, the degree of α is 4.

(c) Show that the degree of β over $Q[\alpha]$ is 2 and hence $\dim_Q Q[\alpha, \beta] = 8$.

(d) Show that the degree of $\alpha + \beta$ is 4.

(e) Find θ such that $Q[\alpha, \beta] = Q[\theta]$.

3. Show that $Q[i, \sqrt{2}, \sqrt{3}] = Q[i + \sqrt{2} + \sqrt{3}]$.

CHAPTER VIII

RINGS OF ALGEBRAIC INTEGERS

8.1 ALGEBRAIC INTEGERS

Questions about the field Q are usually answered by using properties of Z, because Z is highly structured (whereas Q is not) and its structure is understandable. The structure of Z is caused by the presence of the primes; there are no primes in Q since every nonzero element is a unit. The simplicity of Z is provided by the fundamental theorem of arithmetic. It is reasonable to try to study algebraic number fields along the same lines. But what corresponds to Z?

DEFINITION. A number α is said to be an *algebraic integer* if α is a root of a monic polynomial with integer coefficients.

If we did not require that the polynomial be monic, we would just have another definition of an algebraic number (why?). However, the reasons for defining algebraic integers in this particular manner are not too clear at this point. The facts will justify the definition.

THEOREM 47. If $\alpha \in Q$ is an algebraic integer, then α is an integer.

Proof: Let s and t be integers such that $(s, t) = 1$ and $\alpha = s/t$. Suppose α is a root of the polynomial

$$f(x) = x^n + a_{n-1}x^{n-1} + \cdots + a_1 x + a_0,$$

where the a_i are integers. Then

$$(s/t)^n + a_{n-1}(s/t)^{n-1} + \cdots + a_1(s/t) + a_0 = 0$$

so, multiplying through by t^n, we get

$$a_{n-1}s^{n-1}t + a_{n-2}s^{n-2}t^2 + \cdots + a_1 st^{n-1} + a_0 t^n = -s^n.$$

But the left side of this equation is divisible by t; hence $t|s^n$. But $(s, t) = 1$, so $t = \pm 1$. Thus, s/t is an integer.

So the algebraic integers in Q are just the integers; that's a hopeful sign. Theorem 47 can be used to prove that numbers like $\sqrt[3]{7}$ are irrational ($\sqrt[3]{7}$ is a root of $x^3 - 7$, so if it is rational, it would have to be an integer). What are the algebraic integers in the Gaussian numbers, $Q[i]$? If $a + bi$ is a Gaussian integer, then it is a root of the monic polynomial $x^2 - 2ax + (a^2 + b^2)$, which has integer coefficients; thus, Gaussian integers are algebraic integers—another good sign. To show that there are no other algebraic integers in the Gaussian numbers, we characterize those numbers that are algebraic integers in terms of their minimal polynomials.

If the minimal polynomial f of an algebraic number α has *integer* coefficients, then α is an algebraic integer by definition, since α is a root of f. Conversely, if α is an algebraic integer, then f *divides* a monic polynomial with integer coefficients. However, this is enough to guarantee that f itself has integer coefficients.

THEOREM 48. Let g be a monic polynomial with integer coefficients. If $g = hf$ where $h, f \in Q[x]$ and f is monic, then f has integer coefficients.

Proof: Since g and f are monic, so is h. Let s and t be the smallest positive integers such that $sh(x)$ and $tf(x)$ have integer coefficients (s and t are the lowest common denominators of the coefficients of h and f, respectively). Then

$$sh(x) = a_n x^n + a_{n-1}x^{n-1} + \cdots + a_1 x + a_0$$

where $a_n = s$ and, by the minimality of s, the (integer) coefficients of $sh(x)$ have no common prime factors. Similarly, the (integer) coefficients of

$$tf(x) = b_m x^m + b_{m-1}x^{m-1} + \cdots + b_1 x + b_0$$

have no common prime factors. We want to show that $t = 1$, and hence that f has integer coefficients. Suppose that p is a prime factor of t. Then p divides all the coefficients of $stg(x) = (sh(x))(tf(x))$ since g has integer coefficients. But p does not divide all the coefficients of $sh(x)$, so we can let i be the smallest integer such that p does not divide a_i. Similarly, let j be the smallest integer such that p does not divide b_j. Now consider the coefficient of x^{i+j} in $stg(x) = (sh(x))(tf(x))$. It is divisible by p since it is an integer multiple of t. On the other hand, if we multiply out, we find that the coefficient of x^{i+j} is

$$\cdots + a_{i-1}b_{j+1} + a_i b_j + a_{i+1}b_{j-1} + \cdots.$$

Now $a_i b_j$ is not divisible by p since neither a_i nor b_j is. But all the other terms are divisible by p since they contain either a factor a_k with $k < i$, or a factor b_k with $k < j$. Hence, the sum cannot be divisible by p—a contradiction.

COROLLARY 1. Let α be an algebraic number and f its minimal polynomial (over Q). Then α is an algebraic integer if and only if f has integer coefficients.

Proof: If f has integer coefficients, then α is an algebraic integer since $f(\alpha) = 0$ and f is monic. Conversely, if $g(\alpha) = 0$ for some monic polynomial g with integer coefficients, then $g = hf$ for some $h \in Q[x]$, so f has integer coefficients by Theorem 48.

COROLLARY 2. A monic polynomial g with integer coefficients is irreducible in $Q[x]$ if and only if g cannot be written as the product of two nontrivial (that is, not equal to 1) monic polynomials with integer coefficients. Hence, such a polynomial can be written as a product of irreducible monic polynomials with integer coefficients.

Proof: The first part is immediate from Theorem 48. The second part follows upon writing g as a product of monic irreducible polynomials with rational coefficients, and then invoking Theorem 48 to show that these have integer coefficients.

Corollary 2 says that a monic polynomial in $Z[x]$ is irreducible in $Q[x]$ if and only if it is irreducible in $Z[x]$. Notice that we could use Corollary 1 to get a quick proof of Theorem 47: if $\alpha \in Q$, then the minimal polynomial of α is $x - \alpha$, and if $x - \alpha$ has integer coefficients, then α must be an integer! Let's apply it to the Gaussian numbers. If $a + bi$ is a Gaussian number, with $b \neq 0$, then the minimal polynomial of $a + bi$ is $x^2 - 2ax + (a^2 + b^2)$. Hence, $a + bi$ is an algebraic integer if and only if $2a$ and $a^2 + b^2$ are integers. So $a = n/2$ and $a^2 + b^2 = m$ for some integers n and m. Thus, $n^2/4 + b^2 = m$, so $(2b)^2 = 4m - n^2$. Therefore, by Theorem 47, $2b$ is an integer. If we now look at this equation modulo 4, the left side, being a square of an integer, is 0 or 1 depending on whether b is an integer or not. The right-hand side is 0 or -1 depending on whether n is even or odd. Hence, they must both be 0, meaning that b is an integer and $a = n/2$ is an integer. The Gaussian numbers that are algebraic integers are thus precisely the Gaussian integers—a pleasant state of affairs.

If F is an algebraic number field, then the algebraic integers contained in F are as intimately related to F as Z is to Q.

THEOREM 49. If F is an algebraic number field, then any element $\alpha \in F$ can be written in the form β/a where β is an algebraic integer and a is an integer.

Proof: Let $f(x) = ax^n + bx^{n-1} + cx^{n-2} + \cdots$ be a polynomial with integer coefficients that has α as a root, $a \neq 0$; such a polynomial can be obtained by multiplying the minimal polynomial of α by a suitable integer. Then

$$a\alpha^n + b\alpha^{n-1} + c\alpha^{n-2} + \cdots = 0,$$

so, multiplying by a^{n-1}, we have

$$(a\alpha)^n + b(a\alpha)^{n-1} + ca(a\alpha)^{n-2} + \cdots = 0$$

so $a\alpha$ is a root of the monic polynomial $g(x) = x^n + bx^{n-1} + cax^{n-2} + \cdots$. Hence, $\beta = a\alpha$ is an algebraic integer.

8.1 PROBLEMS

1. Show that any integer is an algebraic integer.
2. Let q be a positive rational number. Show that $\sqrt{q}$ is an algebraic integer if and only if q is an integer.
3. Show that $(1 + i\sqrt{3})/2$ is an algebraic integer but $(1+i\sqrt{5})/2$ is not.
4. Show that the product of an algebraic integer and an integer is an algebraic integer.
5. Show that the sum of an algebraic integer and an integer is an algebraic integer.
6. What can you say about the minimal polynomial of an algebraic number α if both α and $1/\alpha$ are algebraic integers? (Note that if α is a root of $a_0 + a_1x + \cdots + a_nx^n$, then $1/\alpha$ is a root of $a_0x^n + a_1x^{n-1} + \cdots + a_n$.)
7. A polynomial f with integer coefficients is said to be *primitive* if its coefficients have no common prime factor.
   (a) Show that any nonzero polynomial g with rational coefficients can be written uniquely as $cf(x)$, where f is primitive and c is a positive rational number. The number c is called the *content* of g. Write the polynomials $3x - 1/2$, $-(2/3)x^2 - (1/4)x - 1/6$, and $(1/5)x^2 - (1/3)x + 1/2$ in this form.

(b) Show that the product of two primitive polynomials is primitive. (Let p be a common prime factor of the coefficients of the product and proceed as in the proof of Theorem 48.)

(c) Show that the content of the product of two polynomials is equal to the product of their contents.

(d) Show that a polynomial with rational coefficients has integer coefficients if and only if its content is an integer. Show that the content of a monic polynomial with rational coefficients is of the form $1/n$, $n \in Z$. Use these results to prove Theorem 48.

8.2 QUADRATIC FIELDS

In both of the fields we have examined, Q and $Q[i]$, the algebraic integers form a ring. A proof that this holds for any algebraic number field is outlined in Problem 6 of this section. We shall content ourselves to look at integers in *quadratic fields*, fields of dimension 2 over Q.

If F is a quadratic field, then $F = Q[\alpha]$ for *any* $\alpha \in F$ that is not in Q. We do not need Theorem 46 here. If $\alpha \in F$ is not rational, then the elements 1 and α are linearly independent over Q (why?). If $\beta \in F$, then, since the dimension of F over Q is 2, the elements 1, α, β cannot be linearly independent. Hence, there are rational numbers a, b, c, not all zero, such that $a + b\alpha + c\beta = 0$. Since 1 and α are linearly independent, we must have $c \neq 0$. Hence, $\beta = -a/c - (b/c)\alpha \in Q[\alpha]$.

We would like to pick a "nice" α. Call an integer *square-free* if it has no nontrivial square factors. For example, 3, 6, 35, and -26 are square-free, whereas $18 = 3^2 \cdot 2$ and $-75 = 5^2(-3)$ are not. Note that any integer is a product of a square-free integer and a square.

THEOREM 50. If F is a quadratic field, then $F = Q[\sqrt{n}]$ for some unique square-free integer n.

Proof: Notice that the choice of sign for $\sqrt{n}$ is irrelevant since $Q[\sqrt{n}] = Q[-\sqrt{n}]$. Suppose $F = Q[\alpha]$. Then, since F is quadratic, α must have a degree 2 (why?). Hence, there are rational numbers b and c such that $\alpha^2 + b\alpha + c = 0$. Thus, $\alpha = (-b + \beta)/2$ where

$\beta^2 = b^2 - 4c$. Since $\beta = 2\alpha + b$, we have $F = Q[\alpha] = Q[\beta]$. Let $b^2 - 4c = s/t$ where s and t are integers. Since $\beta^2 = s/t$, we have $(t\beta)^2 = st$. But $Q[\beta] = Q[t\beta]$, so $F = Q[\sqrt{st}]$. Let $st = a^2 n$ where n is square-free. Then $F = Q[\sqrt{a^2 n}] = Q[a\sqrt{n}] = Q[\sqrt{n}]$.

To show that n is unique, suppose $Q[\sqrt{n}] = Q[\sqrt{m}]$, where m and n are square-free. Then, since $\sqrt{n} \in Q[\sqrt{m}]$, we have $\sqrt{n} = a + b\sqrt{m}$. Squaring both sides, we get $n = a^2 + 2ab\sqrt{m} + b^2 m$. If $ab \neq 0$, we have $\sqrt{m} = (n - a^2 - b^2 m)/2ab \in Q$, which is impossible since $\sqrt{m}$ is irrational. Thus, $ab = 0$ and so $a = 0$ or $b = 0$. If $b = 0$, then $\sqrt{n} = a \in Q$, which is impossible (why?). Thus, $a = 0$ and $\sqrt{n} = b\sqrt{m}$, or $n = b^2 m$. Let $b = s/t$ where s and t are integers such that $(s, t) = 1$. Then $t^2 n = s^2 m$ and, since $(s, t) = 1$, $s^2 \mid n$ and $t^2 \mid m$. But n and m are square-free. Hence, $s^2 = t^2 = 1$, so $n = m$.

We see that the quadratic fields are precisely $Q[\sqrt{2}]$, $Q[\sqrt{3}]$, $Q[\sqrt{5}]$, $Q[\sqrt{6}]$, $Q[\sqrt{7}]$, $Q[\sqrt{10}]$, ... $Q[\sqrt{-1}]$, $Q[\sqrt{-2}]$, $Q[\sqrt{-3}]$, $Q[\sqrt{-5}]$, $Q[\sqrt{-6}]$, ... We have looked only at $Q[\sqrt{-1}]$ so far. What are the algebraic integers in them? If d is a square-free integer, then $\sqrt{d}$ is an algebraic integer; in fact, $a + b\sqrt{d}$ is an algebraic integer for all integers a and b. To see this, we note that $a + b\sqrt{d}$ is a root of the polynomial $x^2 - 2ax + (a^2 - b^2 d)$. Are these the only ones? That was the case for $d = -1$, the Gaussian numbers. However, the roots of $x^2 + x + 1$ are $(-1 \pm \sqrt{-3})/2$, so there are algebraic integers in $Q[\sqrt{-3}]$ which are not of that form. The distinction between these two fields is our old friend, the residue class of d modulo 4.

THEOREM 51. Let d be a square-free integer. Then the algebraic integers in the field $Q[\sqrt{d}]$ are precisely those numbers of the form

(1) $s + t\sqrt{d}$, where $s, t \in Z$, if $d \not\equiv 1 \pmod 4$
(2) $(a + b\sqrt{d})/2$, where $a, b \in Z$ and $a + b$ is even, if $d \equiv 1 \pmod 4$.

Proof: Notice that $d \not\equiv 0 \pmod 4$ since d is square-free (4 is a square). Any number α in $Q[\sqrt{d}]$ can be written in the rather perverse form $\alpha = (a + b\sqrt{d})/2$, where a and b are rational numbers. If $b \neq 0$, the only nontrivial case, then the minimal polynomial of α is $x^2 - ax + (a^2 - b^2d)/4$. Thus, α is an algebraic integer if and only if a is an integer and $(a^2 - b^2d)/4 = n$ is an integer. But the latter equation can be written as $b^2d = a^2 - 4n$, so b must be an integer, for otherwise d, being square-free, would not be able to clear the denominator of b^2. To summarize our results so far, we have shown that $(a + b\sqrt{d})/2$ is an algebraic integer if and only if a and b are integers and $b^2d \equiv a^2 \pmod 4$. Now if $d \equiv 1 \pmod 4$, this happens exactly when $b^2 \equiv a^2 \pmod 4$, which says simply that a and b are either both even or both odd; in other words, $a + b$ is even. If $d \equiv 2$ or $3 \pmod 4$, then $b^2d \not\equiv 1 \pmod 4$, whereas $a^2 \equiv 0$ or $1 \pmod 4$. Hence, $b^2d \equiv a^2 \equiv 0 \pmod 4$, so a and b must be even—that is $\alpha = (a + b\sqrt{d})/2 = s + t\sqrt{d}$, where $s, t \in Z$.

COROLLARY. The algebraic integers in a quadratic field form a ring.

Proof: If $d \not\equiv 1 \pmod 4$, this is clear. If $d \equiv 1 \pmod 4$, the only problem is showing that the product of two algebraic integers is an algebraic integer. We compute:

$$\frac{1}{2}(a + b\sqrt{d}) \cdot \frac{1}{2}(s + t\sqrt{d}) = \frac{1}{2}\left(\frac{1}{2}(as + btd) + \frac{1}{2}(at + bs)\sqrt{d}\right).$$

Now $a + b$ and $s + t$ are even, so $a \equiv b \pmod 2$ and $s \equiv t \pmod 2$. Since $d \equiv 1 \pmod 4$, $d \equiv 1 \pmod 2$ and so $as \equiv btd \pmod 2$ and $at \equiv bs \pmod 2$, that is, $as + btd$ and $at + bs$ are even. Also $(a + b)(s + t) \equiv 0 \pmod 4$, so

$$\begin{aligned} as + btd + at + bs &\equiv as + at + bs + bt \\ &= (a + b)(s + t) \equiv 0 \pmod 4. \end{aligned}$$

Thus, $(as + btd)/2$ and $(at + bs)/2$ are integers, and $(as + btd)/2 + (at + bs)/2$ is even.

In order to prove things about Gaussian integers, we introduced the norm $N(a + bi) = a^2 + b^2$. We generalize this notion to arbitrary quadratic fields.

DEFINITION. If $\alpha = a + b\sqrt{d}$, where a and b are rational numbers and d is a square-free integer, define the *norm* of α by $N(\alpha) = a^2 - b^2 d$, and the *conjugate* of α by $\bar{\alpha} = a - b\sqrt{d}$.

Notice that this definition agrees with the definition of the norm for Gaussian numbers, and if α is irrational, $N(\alpha)$ is the constant term in the minimal polynomial of α. The following theorem collects a few facts about norms and conjugates.

THEOREM 52. Let $\alpha, \beta \in Q[\sqrt{d}]$. Then

(1) $\bar{\alpha} \in Q[\sqrt{d}], \quad \alpha + \bar{\alpha} \in Q, \quad \alpha\bar{\alpha} - N(\alpha) \in Q.$
(2) If $\alpha \notin Q$, then $(x - \alpha)(x - \bar{\alpha}) = x^2 - (\alpha + \bar{\alpha})x + \alpha\bar{\alpha}$ is the minimal polynomial of both α and $\bar{\alpha}$.
(3) $\bar{\alpha}$ is an algebraic integer if and only if α is.
(4) $\overline{\alpha\beta} = \bar{\alpha}\bar{\beta}$.
(5) $N(\alpha\beta) = N(\alpha)N(\beta)$.

Proof: Part 1 is immediate. The polynomial $x^2 - (\alpha + \bar{\alpha})x + \alpha\bar{\alpha}$ certainly has α and $\bar{\alpha}$ as roots and has rational coefficients by part 1; if $\alpha \notin Q$, then α cannot be a root of a polynomial of smaller degree than 2. Part 3 follows from part 2 and Theorem 48. To check part 4, let $\alpha = a + b\sqrt{d}$ and $\beta = s + t\sqrt{d}$. Then $\alpha\beta = as + btd + (at + bs)\sqrt{d}$, so

$$\overline{\alpha\beta} = as + btd - (at + bs)\sqrt{d} = (a - b\sqrt{d})(s - t\sqrt{d}) = \bar{\alpha}\bar{\beta}.$$

Part 5 follows from parts 1 and 4 by

$$N(\alpha\beta) = (\alpha\beta)(\overline{\alpha\beta}) = \alpha\beta\bar{\alpha}\bar{\beta} = (\alpha\bar{\alpha})(\beta\bar{\beta}) = N(\alpha)N(\beta).$$

As in the case of the Gaussian integers, we can tell which algebraic integers in a quadratic field are units just by looking at their norms.

THEOREM 53. Let F be a quadratic number field and R the ring of algebraic integers in F. Then $\alpha \in R$ is a unit if and only if $N(\alpha) = \pm 1$.

Proof: If $\alpha \in Q$, then $\alpha \in Z$ by Theorem 47, and the theorem is clear since $N(\alpha) = \alpha^2$ for $\alpha \in Q$. So we may assume $\alpha \notin Q$. Since F is a field, $1/\alpha \in F$, so α is a unit in R if and only if $1/\alpha$ is an algebraic integer. The minimal polynomial of α is $x^2 + nx + N(\alpha)$ for some $n \in Z$. Thus, $\alpha^2 + n\alpha + N(\alpha) = 0$, so, multiplying through by $1/\alpha^2$, we get $1 + n(1/\alpha) + N(\alpha)(1/\alpha)^2 = 0$, which says that $1/\alpha$ is a root of the polynomial $x^2 + (n/N(\alpha))x + 1/N(\alpha)$. If $\alpha \notin Q$, then $1/\alpha \notin Q$, so $x^2 + (n/N(\alpha))x + 1/N(\alpha)$ is the minimal polynomial of $1/\alpha$. Hence, $1/\alpha$ is an algebraic integer if and only if $1/N(\alpha) \in Z$, which happens if and only if $N(\alpha) = \pm 1$.

If F is a quadratic field and α, β is a basis for F over Q, then we say that α, β is an *integral basis* if α and β are algebraic integers and every algebraic integer in F is (uniquely) of the form $s\alpha + t\beta$ where $s, t \in Z$. Offhand, there may not be any integral bases. However, if we take $\alpha = \sqrt{d}$ when $d \not\equiv 1 \pmod 4$ and $\alpha = (1 + \sqrt{d})/2$ when $d \equiv 1 \pmod 4$, then 1, α is an integral basis (see Problem 5).

8.2 PROBLEMS

1. Show that if α is irrational, then the numbers 1 and α are linearly independent over Q.
2. Show that if f is a ring isomorphism between two algebraic number fields, then $f(q) = q$ for all rational numbers q. (Start with $q = 1$, proceed to q an integer, finish by the relation $f(q)f(m) = f(n)$ for $q = n/m$.)
3. Show that if n and m are distinct square-free integers, then $Q[\sqrt{n}]$ is not ring isomorphic to $Q[\sqrt{m}]$. (If f is an isomorphism from $Q[\sqrt{n}]$ to $Q[\sqrt{m}]$, then $f(\sqrt{n})^2 = n$, which is impossible by the uniqueness part of the proof of Theorem 50.) Are $Q[\sqrt{n}]$ and $Q[\sqrt{m}]$ isomorphic as additive abelian groups?
4. Let F be a quadratic field and R the ring of algebraic integers in F. Show that the function $f(\alpha) = \bar{\alpha}$ defines a ring isomorphism from F to F and a ring isomorphism from R to R. Show that if g is a ring isomorphism from F to F, then either

$g(\alpha) = \alpha$ for all $\alpha \in F$ or $g(\alpha) = \bar{\alpha}$ for all $\alpha \in F$ (α and $g(\alpha)$ must have the same minimal polynomial).

5. Show that the additive group of algebraic integers in a quadratic field is isomorphic to $Z \oplus Z$ (show that every algebraic integer is uniquely of the form $s + t\alpha$ where $\alpha = \sqrt{d}$ if $d \not\equiv 1 \pmod 4$, $\alpha = (1 + \sqrt{d})/2$ if $d \equiv 1 \pmod 4$, and $s, t \in Z$; define $f(s + t\alpha) = (s, t)$).

6. A complex number α is said to be *integral* over a ring R of complex numbers if α satisfies a *monic* polynomial with coefficients in R.
   (a) Show that α is integral over Z if and only if α is an algebraic integer; show that α is integral over Q if and only if α is an algebraic number.
   (b) Show that if α is integral over R, then there is a finite set of nonzero complex numbers $v_1, v_2, \ldots, v_n$ such that αv_i is a linear combination of $v_1, v_2, \ldots, v_n$ *with coefficients in* R for $i = 1, 2, \ldots, n$ (take $1, \alpha, \alpha^2, \ldots, \alpha^{n-1}$ for $v_1, v_2, \ldots, v_n$).
   (c) Show, conversely, that if there exist v_i as in part (b) then α is integral over R. (If $\alpha v_i = \sum a_{ij} v_j$, $a_{ij} \in R$, then $\sum d_{ij} v_j = 0$ where $d_{ij} = -a_{ij}$ if $i \neq j$, and $d_{ii} = \alpha - a_{ii}$. The determinant of d_{ij} is 0, yielding a monic polynomial with coefficients in R satisfied by α.)
   (d) Show that if α and β are integral over R, then so are $\alpha + \beta$, $\alpha - \beta$, and $\alpha\beta$. (Let $v_1, v_2, \ldots, v_n$ and $w_1, w_2, \ldots, w_m$ be associated with α and β as in (b) and (c), and show that the set of products $v_i w_j$ works for $\alpha + \beta$, $\alpha - \beta$ and $\alpha\beta$.) Conclude that the algebraic integers in an algebraic number field form a ring.

8.3 UNITS

If F is a quadratic field and R is the ring of algebraic integers in F, then we shall refer to the units of R simply as the *units of* F. This is taking liberty with the language since, strictly speaking, the units of F are the nonzero elements of F, because F is a field. However, the usage is standard, convenient, and should cause no confusion. The idea is that if you are dealing with a ring, where there are apt to be many elements without inverses, it is reasonable to call those elements that have inverses *units*; if you are dealing with a field, where every nonzero element has an inverse, it is reasonable to reserve the word "unit" for a concept that cuts deeper than the difference between zero and nonzero.

If $d \not\equiv 1 \pmod 4$ is a square-free integer, then $\alpha = a + b\sqrt{d}$ is a unit in $Q[\sqrt{d}]$ if and only if $a, b \in Z$ and $N(\alpha) = a^2 - b^2 d = \pm 1$. Thus, to determine the units of $Q[\sqrt{d}]$ is the same as to find integer solutions a, b to the equation

$$a^2 - b^2 d = \pm 1. \tag{1}$$

If $d \equiv 1 \pmod 4$, then $\alpha = (a + b\sqrt{d})/2$ is a unit in $Q[\sqrt{d}]$ if and only if $a, b \in Z$, $a + b$ is even, and $N(\alpha) = (a^2 - b^2 d)/4 = \pm 1$. Thus, to determine the units of $Q[\sqrt{d}]$ is the same as to find integer solutions a, b to the equation

$$a^2 - b^2 d = \pm 4. \tag{2}$$

We don't have to add the restriction that $a + b$ is even because, if we look at equation 2, modulo 4, we see that this is automatically satisfied by any solution a, b.

Generally, there will be many solutions to equations 1 and 2. However, if $d < 0$, there cannot be too many.

THEOREM 54. Let d be a square-free negative integer. Then the units in $Q[\sqrt{d}]$ are

(1) ± 1 and $\pm i$, if $d = -1$
(2) ± 1 and $(\pm 1 \pm \sqrt{-3})/2$, if $d = -3$
(3) ± 1, if $d \neq -1, -3$.

Proof: We have already proved (1). The units in $Q[\sqrt{-3}]$ are of the form $(a + b\sqrt{-3})/2$, where $a, b \in Z$ and $a^2 + 3b^2 = \pm 4$. Clearly, the only solutions are $a = \pm 2$, $b = 0$, and $a = \pm 1$, $b = \pm 1$, which correspond to the numbers itemized in (2). If $d = -2 \not\equiv 1 \pmod 4$, then we must solve the equation $a^2 + 2b^2 = \pm 1$, which has only the solutions $a = \pm 1$, $b = 0$. If $d < -3$, then $d \leq -5$, so $a^2 - b^2 d \geq 5$ if $b \neq 0$, ruling out any solutions except $a = \pm 1, b = 0$ if $d \not\equiv 1 \pmod 4$ and $a = \pm 2, b = 0$ if $d \equiv 1 \pmod 4$. In either case, these solutions correspond to the units $\pm 1 \in Q[\sqrt{d}]$.

While the units are pretty sparse in $Q[\sqrt{d}]$ when d is negative, the situation is quite different when d is positive. Consider $Q[\sqrt{2}]$. Since $2 \not\equiv 1 \pmod 4$, the algebraic integers look like $a + b\sqrt{2}$ where a and b are integers. The number $a + b\sqrt{2}$ is a unit if and only if

$$a^2 - 2b^2 = \pm 1. \tag{3}$$

The problem of finding the units in $Q[\sqrt{2}]$ is the same as finding all integer solutions to equation 3. You can get a few by trial and error: $a = 1, b = 1$, corresponding to the unit $1 + \sqrt{2}$; $a = 3, b = 2$, corresponding to $3 + 2\sqrt{2}$ etc. Finding *all* integer solutions is another matter. Notice that the essence of the problem is that the solutions must be integers, for otherwise we could simply choose b to be any real number and let $a = \pm\sqrt{1 + 2b^2}$, or if $b \geq \sqrt{2}/2$, $a = \pm\sqrt{2b^2 - 1}$, and these are all the (real) solutions. Equations for which we require integer solutions are called *Diophantine equations.* Complete solutions of Diophantine equations are generally rather involved.

One way to get a whole bunch of solutions to equation 3 is provided by our knowledge that the solutions correspond to units in $Q[\sqrt{2}]$. We found that $\alpha = 1 + \sqrt{2}$ is a unit in $Q[\sqrt{2}]$; it follows that, since the product of units is a unit, so are $\alpha^2, \alpha^3, \alpha^4, \ldots$ (that is, the numbers $3 + 2\sqrt{2}$, $7 + 5\sqrt{2}$, $17 + 12\sqrt{2}, \ldots$). Also $\alpha^{-1} = -1 + \sqrt{2}$ must be a unit and so also must $\alpha^{-2} = 3 - 2\sqrt{2}$, $\alpha^{-3} = -7 + 5\sqrt{2}$, etc. Finally, if β is a unit so is $-\beta$, so the numbers $-1 - \sqrt{2}$, $-3 - 2\sqrt{2}$, $-7 - 5\sqrt{2}, \ldots$ and $1 - \sqrt{2}$, $-3 + 2\sqrt{2}$, $7 - 5\sqrt{2}, \ldots$ are all units.

Let's get organized. If $\beta \neq \pm 1$ is a unit in $Q[\sqrt{2}]$, then so are β^{-1}, $-\beta$ and $-\beta^{-1}$. Of the numbers β, β^{-1}, $-\beta$, and $-\beta^{-1}$, precisely one is bigger than 1 (why?). Moreover, we can get from any of the numbers β, β^{-1}, $-\beta$ and $-\beta^{-1}$ to any other by the processes of taking negatives and reciprocals. So we might as well concentrate on those units that are bigger than 1, because if we know what they are, then we know all.

If $\alpha > 1$ is a unit then $1 < \alpha < \alpha^2 < \alpha^3 < \ldots$ are all units. Moreover, knowledge of α is better than knowledge of, say, α^2 because if we know that α is a unit, then we know that α^2 is a unit, but not conversely. Hence, we want to look for units α that are greater than 1, but not by much. Ideally, we would like the *smallest* unit α that is greater than 1. Offhand, there might not be one—that is, there may be units arbitrarily close to 1. Such is not the case.

LEMMA. There is no unit in $Q[\sqrt{2}]$ between 1 and $1 + \sqrt{2}$.

Proof: Suppose $\alpha = a + b\sqrt{2}$ is a unit in $Q[\sqrt{2}]$ and $1 < \alpha < 1 + \sqrt{2}$. Then $(a - b\sqrt{2})(a + b\sqrt{2}) = a^2 - 2b^2 = N(\alpha) = \pm 1$. Since $a + b\sqrt{2} = \alpha > 1$, we have $|a - b\sqrt{2}| < 1$, so $-1 < a - b\sqrt{2} < 1$. Adding this last inequality to $1 < a + b\sqrt{2} < 1 + \sqrt{2}$, we get $0 < 2a < 2 + \sqrt{2}$, and so $0 < a < 1 + \sqrt{2}/2$. Since a is an integer, we must have $a = 1$. But there is no integer b satisfying $1 < 1 + b\sqrt{2} < 1 + \sqrt{2}$.

This lemma enables us to determine all the units in $Q[\sqrt{2}]$.

THEOREM 55. The units of $Q[\sqrt{2}]$ are the numbers of the form $\pm(1 + \sqrt{2})^n$, where n is an integer.

Proof: By the argument preceding the lemma, it suffices to show that every unit bigger than 1 is of the form $(1 + \sqrt{2})^n$ for some positive integer n. Let $\alpha > 1$ be a unit. Then there exists a positive integer n such that $(1 + \sqrt{2})^n \leq \alpha < (1 + \sqrt{2})^{n+1}$, since we can make $(1 + \sqrt{2})^n$ as large as we please and $1 + \sqrt{2} \leq \alpha$. Multiplying through by $(1 + \sqrt{2})^{-n}$ we get $1 \leq \alpha(1 + \sqrt{2})^{-n} < 1 + \sqrt{2}$. Since $\alpha(1 + \sqrt{2})^{-n}$ is a unit, then, by the lemma, we cannot have $1 < \alpha(1 + \sqrt{2})^{-n} < 1 + \sqrt{2}$. Hence, $1 = \alpha(1 + \sqrt{2})^{-n}$, so $\alpha = (1 + \sqrt{2})^n$.

8.3 PROBLEMS

1. Show that if β is a nonzero real number such that $\beta \neq \pm 1$, then the numbers β, β^{-1}, $-\beta$, and $-\beta^{-1}$ are all distinct and precisely one of them is greater than 1.

2. Show that the group of units of $Q[\sqrt{2}]$ is isomorphic to $Z_2 \oplus Z$.

3. Show that if α is a unit in $Q[\sqrt{3}]$, then $N(\alpha) = 1$. (Look at the equation $a^2 - 3b^2 = \pm 1$, modulo 4.)

4. Show that there is no unit in $Q[\sqrt{3}]$ between 1 and $2 + \sqrt{3}$ (proceed as in the proof of the lemma). Conclude that the units of $Q[\sqrt{3}]$ are the numbers of the form $\pm (2 + \sqrt{3})^n$ for n an integer.

5. Show that if d is a negative integer, then the units in $Q[\sqrt{d}]$ form a cyclic group.

8.4 UNIQUE FACTORIZATION

The problem of determining the primes in the ring of algebraic integers of an arbitrary quadratic field is more complicated than it was for the Gaussian integers. It is also less rewarding, and for the same reason: in general, we do not have unique factorization into primes.

Consider the algebraic integers in $Q[\sqrt{-5}]$. Since $5 \not\equiv 1 \pmod 4$, they look like $a + b\sqrt{-5}$, $a, b \in Z$. We know that the units are just ± 1 by Theorem 54. The number 2 is prime because $N(2) = 4$, and if $\alpha\beta = 2$, then $N(\alpha)N(\beta) = 4$ and hence, if neither α nor β is a unit, then $N(\alpha) = N(\beta) = 2$. But $N(a + b\sqrt{-5}) = a^2 + 5b^2$, and there are no integers a and b that make $a^2 + 5b^2 = 2$, as you can easily check. In fact, this argument shows that if $N(\alpha) = 4$, then α is a prime, hardly a stirring generalization since the only algebraic integers in $Q[\sqrt{-5}]$ of norm 4 are 2 and -2.

Similarly, 3 is a prime since $N(3) = 9$, and there are no integers a and b such that $a^2 + 5b^2 = 3$. But there are other algebraic integers in $Q[\sqrt{-5}]$ of norm 9, such as $2 + \sqrt{-5}$ and $2 - \sqrt{-5}$. So these numbers are also prime. But then $3 \cdot 3 = 9 = (2 + \sqrt{-5})(2 - \sqrt{-5})$ gives us two distinct ways of writing 9 as a product of primes.

This phenomenon is not confined to $Q[\sqrt{-5}]$. Consider $Q[\sqrt{10}]$. We can write $10 = \sqrt{10} \cdot \sqrt{10} = 2 \cdot 5$. Since $N(\sqrt{10}) = -10$, $N(2) = 4$, and $N(5) = 25$, a nontrivial factorization of $\sqrt{10}$, 2, or 5 would involve an algebraic integer of norm ± 2 or ± 5. However, if $a^2 - 10b^2 = \pm 2$ for integers a and b, then $a^2 \equiv \pm 2 \pmod{10}$, which cannot be true, as you can determine by checking the ten residue classes in Z_{10}. Also, if $a^2 - 10b^2$

$= \pm 5$, then a^2, and hence a, is divisible by 5, say $a = 5c$. Thus, $25c^2 - 10b^2 = \pm 5$, and so $5c^2 - 2b^2 = \pm 1$. Look at this last equation modulo 5. We get $-2b^2 \equiv \pm 1 \equiv \mp 4 \pmod 5$, so $b^2 \equiv \pm 2 \pmod 5$, which cannot be true (check in Z_5). So $\sqrt{10}$, 2, and 5 are all primes and, since the absolute values of their norms are different, none is a unit multiple of another.

This is a nasty situation: if we do not have unique factorization into primes, then we cannot extend our ordinary arithmetic techniques to number fields. The trouble turns out to be that we are in Plato's cave; there are things we cannot see governing the phenomena we observe.

Let's forget about addition for a minute and concentrate on multiplication, which is all that is involved in unique factorization. Suppose we were aware only of the numbers 1, 4, 7, 10, 13, . . . ; that is, the positive integers that are congruent to 1 modulo 3. We can multiply these integers and never get anything new. We have a number 1 with the right properties, and we can talk about factoring and about primes. For example, 4 is a prime because the only ways to write 4 as a product are $4 = 4 \cdot 1 = 1 \cdot 4$. Similarly 10 and 25 are primes. But $10 \cdot 10 = 4 \cdot 25$, so there are two distinct ways to write 100 as a product of primes.

To someone aware of the rest of the integers, this appears foolish. He would say that our problem is simply that we are not looking at *all* the integers, and that the numbers we think are primes are not primes at all. Perhaps the same thing is going on in rings of algebraic integers; if we could find out what the primes *really* are, we would have unique factorization. This approach turns out to be extremely fruitful and leads to one of the most important notions in modern algebra: the ideal.

8.4 PROBLEMS

1. Let R be the ring of algebraic integers in $Q[\sqrt{10}]$. Show that no element of R has norm ± 3. Conclude that $1 + \sqrt{10}$, $1 - \sqrt{10}$, and 3 are primes in R, and hence that 9 can be written as a product of primes in two distinct ways.
2. Let d be a square-free negative integer. Show that 2 is a prime in the ring of algebraic integers in $Q[\sqrt{d}]$ if and only if $d < -3$.
3. Show that unique factorization fails in $Q[\sqrt{-13}]$.

4. Imitate the proof of Theorem 32 to prove a division algorithm for the algebraic integers in $Q[\sqrt{2}]$ using $|N(\alpha)|$ instead of $N(\alpha)$. Conclude that unique factorization holds in $Q[\sqrt{2}]$.

CHAPTER IX

IDEAL THEORY

9.1 IDEALS

Recall the plight of the people who only had the numbers 1, 4, 7, 10, 13, ... at their disposal. They thought that 100 could be written in two distinct ways as a product of primes, as $10 \cdot 10$ and $4 \cdot 25$. We see that the problem is their ignorance of the numbers 2 and 5. If we tried to tell them that 4 is not a prime because it is divisible by 2, we would be greeted by blank stares, since the number 2 has no meaning to them. We have to relate 2 to their experience. One way is to make use of the correspondence between a number n and the set of all multiples of n, the set of numbers that n divides. We can't tell them what 2 "is," but we can tell them what 2 "does."

We might try to communicate the notion of 2 by pointing to the subset 4, 10, 16, 22, 28, ... and informing them that these are precisely the numbers (that they know of) that are multiples of 2. They might, if they believed us, come to think of 2 as being some sort of "ideal" number that divided the numbers in this subset, and no others. If they wanted to think of something concrete, they could always just think of that subset.

If we attempt to apply this idea to a ring R of algebraic integers, we look for subsets of R that might possibly be the set of all multiples of something. The obvious properties of a set I of multiples of something are

(1) $0 \in I$.
(2) If $x, y \in I$, then $x + y \in I$.
(3) If $x \in I$ and $r \in R$, then $rx \in I$.

But these are just the defining conditions for an *ideal* in a ring (see section 1.2). For the ring of integers, we showed that every ideal was, in fact, the set of multiples of some integer (Theorem 3). We can view this as meaning that there are no "ideal" integers that we can discover by looking for the set of integers that they divide. And we don't need any ideal integers, because of the fundamental theorem of arithmetic. However, in rings of algebraic integers, we would like to do something to get unique factorization; ideals turn out to be the answer.

Let's get some notation so we can talk efficiently about ideals. If $a \in R$, we denote by (a) the set of all elements of the form ra, $r \in R$. It is easily seen that (a) is an ideal; we call (a) the *ideal generated by* a, and any ideal I of the form $I = (a)$ is called a *principal ideal* with *generator* a. The ring R is a principal ideal with generator 1; any other ideal of R is said to be *proper* (as in "proper subset"). The set $\{0\}$ is a principal ideal with generator 0; it is referred to as the *zero ideal*. More generally, if $a_1, a_2, \ldots, a_n \in R$, we denote by $(a_1, a_2, \ldots, a_n)$ the set of all elements of the form $r_1a_1 + r_2a_2 + \cdots + r_na_n$ where $r_1, r_2, \ldots, r_n \in R$. This is clearly an ideal and, in fact, the smallest ideal containing $a_1, a_2, \ldots, a_n$; we call it the *ideal generated by* $a_1, a_2, \ldots, a_n$.

DEFINITION. If A and B are ideals in the ring R, then the *product* AB is defined to be the set of all sums of the form $a_1b_1 + a_2b_2 + \cdots a_nb_n$ where the a's are in A and the b's are in B.

THEOREM 56. Let A, B, and C be ideals in the ring R. Then

(1) AB is an ideal in R.
(2) $RA = A$.
(3) $AB = BA$.
(4) $(AB)C = A(BC)$.
(5) If $A = (\alpha)$ and $B = (\beta)$, then $AB = (\alpha\beta)$.
(6) If A is generated by α_i, $i = 1, 2, \ldots, n$ and B is generated by β_j, $j = 1, 2, \ldots, m$, then AB is generated by $\alpha_i\beta_j$, $i = 1, 2, \ldots, n$, $j = 1, 2, \ldots, m$.

Proof: To see that AB is an ideal, we check the three properties: $0 \in AB$ because $0 \in A$, $0 \in B$, and $0 = 0 \cdot 0$; it is clear that the sum of two elements of the form given in the definition is again of the same form; and if $r \in R$, then

$$r(a_1b_1 + a_2b_2 + \cdots + a_nb_n) = (ra_1)b_1 + (ra_2)b_2 + \cdots + (ra_n)b_n$$

is of the right form because $ra_i \in A$, since A is an ideal.

Certainly $A \subseteq RA$, for if $a \in A$, then $a = 1 \cdot a$, where $1 \in R$ and $a \in A$. If $a_1, a_2, \ldots, a_n \in A$ and $r_1, r_2, \ldots, r_n \in R$, then

$$r_1a_1 + r_2a_2 + \cdots + r_na_n \in A$$

because A is an ideal. Hence, $RA = A$. Indeed, saying $RA = A$ is the same as saying that A is an ideal.

Part 3 follows from the identity $\sum a_ib_i = \sum b_ia_i$ and part 4 from the fact that either side consists precisely of those elements of the form $\sum a_ib_ic_i$, where $a_i \in A$, $b_i \in B$, and $c_i \in C$.

If $A = (\alpha)$ and $B = (\beta)$, then any element $a_1b_1 + a_2b_2 + \cdots + a_nb_n$ where $a_i \in A$ and $b_i \in B$ can be written

$$(r_1\alpha)(s_1\beta) + (r_2\alpha)(s_2\beta) + \cdots + (r_n\alpha)(s_n\beta) = (r_1s_1 + r_2s_2 + \cdots + r_ns_n)\alpha\beta \in (\alpha\beta),$$

where r_i, $s_i \in R$. Thus, $AB \subseteq (\alpha\beta)$. Since $\alpha\beta \in AB$, we have $AB = (\alpha\beta)$. Part 6 is proved similarly.

Parts 1 through 4 of Theorem 56 say that ideals behave, under multiplication, more or less like numbers, with R playing the role of the number 1. Part 5 says that if we identify a number with the principal ideal that it generates, then multiplication of numbers corresponds to multiplication of ideals.

Let's look at some ideals in the ring of algebraic integers in $Q[\sqrt{-5}]$. Since $3 \cdot 3 = (2 + \sqrt{-5})(2 - \sqrt{-5})$, we might try to find an "ideal number" that divides both 3 and $2 + \sqrt{-5}$. This leads us to consider the ideal $A = (3, 2 + \sqrt{-5})$, all of whose elements would be divisible by this "ideal number." Similarly, we consider the ideal $B = (3, 2 - \sqrt{-5})$. Now $AB = (9, 6 - 3\sqrt{-5}, 6 + 3\sqrt{-5}, 9)$. Every element of AB is divisible by 3 since all the generators are. Hence, $AB \subseteq (3)$. But $3 = (6 - 3\sqrt{-5}) + (6 + 3\sqrt{-5}) - 9 \in AB$, so $AB = (3)$. We have succeeded in factoring the ideal (3) into a product of ideals. Now $A^2 = (9, \; 6 + 3\sqrt{-5}, \; 6 + 3\sqrt{-5}, \; -1 + 4\sqrt{-5}) \subseteq (2 + \sqrt{-5})$, since $9 = (2 - \sqrt{-5})(2 + \sqrt{-5})$, $6 + 3\sqrt{-5} = 3(2 + \sqrt{-5})$, and $-1 + 4\sqrt{-5} = (2 + \sqrt{-5})^2$. On the other hand, $2 + \sqrt{-5} = 9 - (6 + 3\sqrt{-5}) - 1 + 4\sqrt{-5}$ is in A^2. Hence, $A^2 = (2 + \sqrt{-5})$. Similarly, $B^2 = (2 - \sqrt{-5})$. Thus, both sides of the equation $3 \cdot 3 = (2 + \sqrt{-5})(2 - \sqrt{-5})$ can be further factored, yielding $AB \cdot AB = A^2B^2$, and the factorizations are the same.

We may interpret this as follows: we can factor the number 9 into primes in two distinct ways in the ring of algebraic integers in $Q[\sqrt{-5}]$. However, if we allow ideals, then we can factor both sides more and end up with the same factors; we have discovered the "reason" why $3 \cdot 3 = (2 + \sqrt{-5})(2 - \sqrt{-5})$. We still have to ask whether A and B are "primes" and whether A^2B^2 is the only way we can factor the ideal (9) into such "primes."

9.1 PROBLEMS

1. Show that if R is a ring and $a_1, a_2, \ldots, a_n \in R$, then $(a_1, a_2, \ldots, a_n)$ is the smallest ideal in R containing $a_1, a_2, \ldots, a_n$.
2. If x and y are integers, we have defined (x, y) in two ways: as the greatest common divisor of x and y, and as the ideal generated by x and y. How are these two notions related?
3. Prove part 6 of Theorem 56.

4. Let R be the ring of algebraic integers in $Q[\sqrt{10}]$. Let $A = (\sqrt{10}, 2)$ and $B = (\sqrt{10}, 5)$. Compute A^2, B^2, and AB.
5. Show that a ring R is a field if and only if the only ideals of R are R and $\{0\}$.
6. Show that an ideal is an abelian group under addition.
7. Show that if p is a prime number and R is a ring with exactly p elements, then R is a field.
8. Show that if A and B are ideals, then $AB \subseteq A \cap B$.

9.2 PRIME IDEALS

We wish to extend the notion of "prime" to ideals. What properties should we require a prime ideal to have? More concretely, what do we want the prime ideals in Z to be? Since every ideal in Z is principal, the natural candidates are the ideals generated by prime numbers. If we examine the properties of these ideals, perhaps we can capture the essence of "primeness." Let p be a prime number and P the ideal generated by p. Denoting the ring of integers by R for the purpose of generalization, we have that P is a proper ideal and, if A and B are any ideals in R, then

(1) If $P = AB$, then $A = R$ or $B = R$ (so $P = B$ or $P = A$).
(2) If $P \supseteq AB$, then $P \supseteq A$ or $P \supseteq B$.
(3) If $A \supseteq P$, then $A = R$ or $A = P$.

The verification that P has these three properties is left as an exercise. Property 1 seems the most natural since it directly imitates the definition of prime number (the ideal R acts like the number 1 under multiplication). However, it turns out that property 2 is a more fruitful notion in general, although the two properties are equivalent for the rings we are interested in.

DEFINITION. If R is a ring and P is a proper ideal in R, then P is said to be *prime* if whenever $ab \in P$ for $a, b \in R$, then $a \in P$ or $b \in P$.

This definition is actually a disguised version of property 2. You should check that, indeed, a proper ideal P in a ring R is prime if and only if P satisfies property 2.

In Z, the prime ideals are precisely $\{0\}$ and the ideals generated by prime numbers; this follows easily from the fact that every ideal in Z is principal. Let's look at some other prime ideals. Consider the ideal $A = (3, 2 + \sqrt{-5})$ in the ring R of algebraic integers in $Q[\sqrt{-5}]$. In order to check whether A is prime or not, we need a good way of telling when an element $a + b\sqrt{-5}$ is in A. For example, is $1 + \sqrt{-5}$ in A? The secret is to look at $a + b$ modulo 3. If $a + b \equiv 0 \pmod 3$, then $a + b = 3n$, so

$$a + b\sqrt{-5} = a + (3n - a)\sqrt{-5} = a + (3(n-a) + 2a)\sqrt{-5}$$
$$= a(1 + 2\sqrt{-5}) + 3(n - a)\sqrt{-5} \in A,$$

since $1 + 2\sqrt{-5}$ and 3 are in A. Conversely, every element of A has the form

$$(a + b\sqrt{-5})3 + (c + d\sqrt{-5})(1 + 2\sqrt{-5}) = (3a + c - 10d)$$
$$+ (3b + 2c + d)\sqrt{-5}$$

and

$$(3a + c - 10d) + (3b + 2c + d) = 3a + 3b + 3c - 9d \equiv 0 \pmod 3.$$

Thus $a + b\sqrt{-5} \in A$ if and only if $a + b \equiv 0 \pmod 3$.

In particular, $1 \notin A$, so A is proper. Suppose now that the product $(a + b\sqrt{-5})(c + d\sqrt{-5})$ is in A. Then $ac - 5bd + ad + bc \equiv 0 \pmod 3$, so

$$ac + bd + ad + bc = (a + b)(c + d) \equiv 0 \pmod 3$$

since $-5 \equiv 1 \pmod 3$. But Z_3 is a field, so either $a + b \equiv 0 \pmod 3$ or $(c + d) \equiv 0 \pmod 3$, so either $a + b\sqrt{-5} \in A$ or $c + d\sqrt{-5} \in A$; thus, A is a prime ideal. Actually, we can say a little more about A.

DEFINITION. If R is a ring and M is an ideal in R, then M is a *maximal* ideal if $M \neq R$ and, if I is an ideal in R such that $I \supseteq M$, then $I = M$ or $I = R$.

A proper ideal M is maximal if it is contained in no other proper ideals. This is just property 3 of ideals in Z generated by prime numbers. It is a stronger condition than property 2.

THEOREM 57. If R is a ring and M is a maximal ideal in R, then M is prime.

Proof: Suppose $ab \in M$ and $a \notin M$. We shall prove that $b \in M$. Consider the set I of all elements of the form $ra + m$, where $r \in R$ and $m \in M$. It is easy to check that I is an ideal. Certainly $I \supseteq M$ because every element $m \in M$ may be written in the form $0 \cdot a + m$; also, since $a = 1 \cdot a + 0 \in I$ and $a \notin M$, we have $I \neq M$. Hence, since M is maximal, $I = R$; in particular, $1 \in I$, so there are elements $r \in R$ and $m \in M$ such that $ra + m - 1$. Multiplying both sides of this equation by b, we get $rab + mb = b$. But $ab \in M$ and $m \in M$, so $rab + mb = b \in M$.

Every maximal ideal is prime but, in general, not every prime ideal is maximal, even if we exclude the ideal $\{0\}$ (see Problem 4). We shall show that $A = (3, 2 + \sqrt{-5})$ is, in fact, a maximal ideal. Suppose I is an ideal such that $A \subseteq I$ and $A \neq I$. Then there is an element $a_o + b_o\sqrt{-5} \in I$ such that $a_o + b_o \not\equiv 0 \pmod 3$. Now for any integers a and b, either $a + b \equiv 0 \pmod 3$, or $a + b \equiv a_o + b_o \pmod 3$, or $a + b \equiv 2(a_o + b_o) \pmod 3$, since $2 \cdot 1 \equiv 2$ and $2 \cdot 2 \equiv 1 \pmod 3$. Thus, $a + b \equiv n(a_o + b_o) \pmod 3$ for some integer n. But then $a + b\sqrt{-5} - n(a_o + b_o\sqrt{-5}) \in A$, so $a + b\sqrt{-5} = n(a_o + b_o\sqrt{-5}) + \alpha$ for some $\alpha \in A$. Since $n(a_o + b_o\sqrt{-5})$ and α are both in I, so is $a + b\sqrt{-5}$. We have shown that any element $a + b\sqrt{-5}$ in R is in I; hence, $I = R$, so A is maximal.

9.2 PROBLEMS

1. Show that if P is a proper ideal in a ring R, then P is a prime ideal if and only if P satisfies property 2.
2. Show that if p is a prime number and P is the ideal in $R = Z$ generated by p then P satisfies properties 1, 2, and 3.
3. Show that if P is a proper ideal in $R = Z$ and P satisfies property 1, 2, or 3, then $P = \{0\}$ or P is generated by a prime number. Which of the properties 1, 2, and 3 does the ideal $\{0\}$ satisfy?

4. Let $R = Z[x]$, the ring of polynomials in one variable with integer coefficients. Let P be the set of polynomials with constant term 0, and M the set of polynomials with even constant term. Show that M is a maximal ideal and P is a prime ideal but not a maximal ideal.
5. Show that the ideal $A = (2, \sqrt{10})$ is maximal in the ring of algebraic integers in $Q[\sqrt{10}]$ (show that $a + b\sqrt{10} \in A$ if and only if a is even).
6. Show that if P is a prime ideal and $P \supseteq A_1 \cdot A_2 \cdots A_n$, where the A_i are ideals, then $P \supseteq A_i$ for some i.
7. Find the prime ideals in Z_6, Z_8, and $Q \oplus Q$. What are the maximal ideals?

9.3 RESIDUE CLASS RINGS

The notion of congruence of integers modulo n can be expressed in terms of ideals. If we let I be the principal ideal generated by n, then $a \equiv b \pmod{n}$ if and only if $a - b \in I$. This observation allows us to extend the useful notion of congruence to arbitrary rings and ideals.

DEFINITION. If R is a ring, I an ideal in R, and $a, b \in R$, we say that *a is congruent to b modulo I* if $a - b \in I$.

The notion of congruence modulo I is most conveniently thought of in the language of cosets. Recall (in the proof of Lagrange's theorem) that if I is a subgroup of an (additive) abelian group R, then a *coset* of I is a set of elements of the form $a + I$ for some $a \in R$, where $a + I$ consists of all sums $a + i$, where $i \in I$. Every element $a \in R$ is in a unique coset of I, namely $a + I$. Indeed, $a \in a + I$ since $0 \in I$. Moreover, if $a \in b + I$, then $a = b + i_0$ for some $i_0 \in I$, so $a + i = b + (i_0 + i) \in b + I$ for all $i \in I$, which shows that $a + I \subseteq b + I$, while $b + i = a + (i - i_0) \in a + I$ for all $i \in I$, which shows that $b + I \subseteq a + I$. Thus, a and b are in the same coset of I if and only if $a \in b + I$, which is just another way of saying that a is congruent to b modulo I.

Denote the set of cosets of I by R/I. Then R/I is a ring in a natural way. If A and B are in R/I, then all the sums $a + b$, for $a \in A$ and

$b \in B$, lie in the same coset of I. In fact, if $a' = a + i_0$ and $b' = b + i_1$, then

$$a' + b' = a + b + (i_0 + i_1) \in (a + b) + I.$$

Define $A + B$ to be this coset. Similarly, all products ab lie in the same coset of I, for if $a' = a + i_0$ and $b' = b + i_1$ then

$$a'b' = ab + ai_1 + bi_0 + i_1 i_0 \in ab + I.$$

Define AB to be this coset. These definitions can be summarized by

$$\begin{aligned}(a + I) + (b + I) &= (a + b) + I \\ (a + I)(b + I) &= ab + I,\end{aligned}$$

which hold for all elements a and b in R. The fact that R/I is a ring follows easily from the fact that R is a ring. This construction is completely analogous to the construction of Z_n, which is achieved by taking $R = Z$ and $I = (n)$.

Roughly speaking, ideals are the subsets of rings that can be set equal to zero. For example, suppose we want to set $3 = 0$ in Z. We can't stop there and still have the usual rules of arithmetic, for if $3 = 0$, then $3n = 0n = 0$ for any integer n. So we have to set all multiples of 3 equal to 0 also. When we do this, we are essentially left with just the three numbers 0, 1, and 2, since every integer can be written in the form $3n + r$ where $r = 0$, 1, or 2; and $3n = 0$. In this manner, we get Z_3: we start out setting $3 = 0$, and end up setting every element in the ideal (3) equal to 0.

What happens if we want to set some elements in an arbitrary ring equal to zero? We have to keep 0 equal to 0; if we set $a = 0$ and $b = 0$, then certainly we must set $a + b = 0 + 0 = 0$; and if we set $a = 0$, we must set $ra = r0 = 0$. This gives us the three properties of an ideal; an ideal is just a generalization of 0! Moreover, the construction of R/I shows that we *can* set all the elements of an ideal I equal to 0, since the zero of R/I is just $0 + I = I$.

Many features of the ideal I can be interpreted in R/I. It is a good exercise in definitions to prove that an ideal $I \neq R$ is prime if and only if R/I is an integral domain. We also have the following characterization of maximal ideals.

THEOREM 58. If R is a ring and I is a proper ideal in R, then I is maximal if and only if R/I is a field.

Proof: Suppose I is maximal. If $a + I \neq 0$ in R/I, we must have $a \notin I$. Let J be the set of elements in R of the form $ra + i$, where $r \in R$ and $i \in I$. Then J is an ideal, $J \supseteq I$ and, since $a \in J$, $J \neq I$. Hence, $J = R$ and, in particular, $1 \in J$. Thus, there exist $r \in R$ and $i \in I$ such that $ra + i = 1$. But this implies that $(r + I)(a + I) = ra + I = 1 + I$, the "1" of R/I. Hence, $a + I$ has an inverse in R/I.

Conversely, if R/I is a field and J is an ideal containing I but not equal to I, then for $a \in J$ but $a \notin I$, we have $a + I \neq 0$ in R/I, so $ra + I = (r + I)(a + I) = 1 + I$ for some $r \in R$. Thus, $1 \in ra + I$; that is, $1 = ra + i$ for some $i \in I$. But $a \in J$ and $i \in I \subseteq J$, so this says that $1 \in J$, so $J = R$.

Since every field is an integral domain, Theorem 58 together with the "good exercise" gives another proof that every maximal ideal is prime. Conversely, we may use Theorem 11 to show that if I is prime and R/I is finite, then I is maximal. For rings of algebraic integers in quadratic fields we have this subsidiary condition.

THEOREM 59. If $I \neq \{0\}$ is an ideal in the ring R of all algebraic integers in a quadratic field, then R/I is finite. If $I = (n)$ for some nonzero integer n, then R/I has precisely n^2 elements.

Proof: Let $1, \alpha$ be an integral basis for R and n a nonzero element of $I \cap Z$ (for example, $n = N(\alpha)$ for $0 \neq \alpha \in I$). Suppose $s + t\alpha$ is any element of R where $s, t \in Z$. If we write $s = q_1 n + r_1$ and $t = q_2 n + r_2$, where $0 \leq r_1, r_2 < n$, then $s + t\alpha = (q_1 + q_2\alpha)n + (r_1 + r_2\alpha)$ is congruent to $r_1 + r_2\alpha$ modulo I. But there are only n^2 possible choices for the pair of integers r_1, r_2, so there are no more than n^2 distinct cosets of I, meaning that R/I is finite. Moreover, if $I = (n)$, then no two of the numbers $r_1 + r_2\alpha$ are congruent modulo I, so there are precisely n^2 elements in R/I.

COROLLARY. If R is the ring of all algebraic integers in a quadratic field and P is a nonzero prime ideal of R, then P is a maximal ideal.

Proof: Since P is prime, R/P is an integral domain. But by Theorem 59, R/P is finite. Thus, by Theorem 11, R/P is a field; so by Theorem 58, P is a maximal ideal.

9.3 PROBLEMS

1. Prove that if R is a ring and I is an ideal in R, then R/I is a ring.
2. Show that if R is a ring and $I \neq R$ is an ideal in R, then I is prime if and only if R/I is an integral domain.
3. Let $R = Z_{15}$ and I be the ideal in R generated by 5. Show that R/I is a field isomorphic to Z_3.
4. Let R be the ring of Gaussian integers and I the ideal in R generated by 3. Show that R/I is a nine-element field. Is R/I isomorphic to Z_9?
5. Find the maximal ideals in Z_{21} and Z_{25}. What are the corresponding residue class rings?
6. Let $R = \mathscr{R}[x]$, the ring of polynomials with real coefficients. Let I be the principal ideal in R generated by $x^2 + 1$. Show that R/I is isomorphic to the field of complex numbers (use the division algorithm to show that every element α of R/I is the residue class of a unique element of the form $a + bx \in \mathscr{R}[x]$. Define $f(\alpha) = a + bi$).
7. Let $A = (3, 2 + \sqrt{-5})$ in the ring R of algebraic integers in $Q[\sqrt{-5}]$. Show that $R/A \cong Z_3$.
8. Let I be the ideal $(2, \sqrt{10})$ in the ring R of algebraic integers $Q[\sqrt{10}]$. What is the residue class ring R/I (see Problem 5 of the last section)?

9.4 INVERTIBLE IDEALS

Let F be a quadratic field and R the ring of algebraic integers in F. Our ultimate goal is to show that every nonzero ideal in R can be written uniquely as a product of prime ideals. Let's look at the uniqueness part in the simplest situation. Suppose P, Q_1, and Q_2 are prime ideals and $PQ_1 = PQ_2$. We wish to conclude that $Q_1 = Q_2$; that is, we want to "cancel" the P from both sides of the equation. If we were dealing with numbers rather than ideals, we could accomplish this by multiplying both sides by P^{-1}. What happens if we attempt to extend this idea to ideals?

DEFINITION. If A is a nonzero ideal in R, define A^{-1} to be the set of elements $f \in F$ such that $fa \in R$ for every $a \in A$.

Let's examine this notion to see if it is reasonable. Note first that if $A = (a_1, a_2, \ldots, a_n)$, then $f \in A^{-1}$ if and only if $fa_i \in R$ for $i = 1, 2, \ldots, n$; it suffices to check f against the *generators* of A. In particular, if $A = (a)$ is principal, then $f \in A^{-1}$ if and only if $fa \in R$—that is, if $f = r/a$ for some $r \in R$. Thus, A^{-1} consists of all multiples of $1/a$ by elements of R. That is reasonable: the inverse of a principal ideal A is generated by the inverse of the generator of A. Notice that A^{-1} has all the properties of an ideal, except we allow it to be a subset of F rather than just of R. Such an object is called a *fractional ideal.* We can talk about a fractional ideal B being generated by a set of elements $b_1, b_2, \ldots, b_n$ in F: we mean that B is the set of all elements in F of the form $r_1b_1 + r_2b_2 + \cdots + r_nb_n$ where the r_i are in R. As for ideals, we write $B = (b_1, b_2, \ldots, b_n)$. In this notation, if $A = (a)$, then $A^{-1} = (a^{-1})$.

An ideal is a fractional ideal that is contained in R. The product of fractional ideals is defined in the same way as the product of ideals, and you can check that all of the properties in Theorem 56 hold for fractional ideals as well as ideals. An ideal A is said to be *invertible* if $A^{-1}A = R$. Note that if A is any ideal, then $A^{-1} \supseteq R$ and $A \subseteq A^{-1}A \subseteq R$, so the problem is to show that $A^{-1}A \supseteq R$—that is, that $1 \in A^{-1}A$. Note also that if $A = (a)$ is a principal ideal, then $A^{-1}A = (a^{-1})(a) = (a^{-1}a) = (1) = R$—every nonzero principal ideal is invertible. Indeed, invertible ideals may be thought of as the next best thing to principal ideals. In Z, every ideal is principal; in the ring of algebraic integers in a quadratic field, every nonzero ideal is invertible.

Returning to our problem of canceling P in the equation $PQ_1 = PQ_2$, we see that there is no problem if P is invertible, for we can multiply both sides by P^{-1} getting $Q_1 = RQ_1 = P^{-1}PQ_1 = P^{-1}PQ_2 = RQ_2 = Q_2$, as if we were dealing with numbers.

Let's look again at the ideal $A = (3, 2 + \sqrt{-5})$ in the ring of algebraic integers in $Q[\sqrt{-5}]$. Now $1 \in A^{-1}$ for any ideal A. Another element of A^{-1} is $(2 - \sqrt{-5})/3$. Indeed, $(2 - \sqrt{-5})/3 \cdot 3 = 2 - \sqrt{-5} \in R$ and $(2 - \sqrt{-5})/3 \cdot (2 + \sqrt{-5}) = 9/3 = 3 \in R$. But

$$1 = (2 - \sqrt{-5}) + (2 + \sqrt{-5}) - 3 = (2 - \sqrt{-5})/3 \cdot 3 + 1 \cdot ((2 + \sqrt{-5}) - 3) \in A^{-1}A,$$

so A is invertible. We use exactly the same procedure to show that any nonzero ideal in the ring of algebraic integers in a quadratic field is invertible. First we need a "nice" representation like the one we have for $(3, 2 + \sqrt{-5})$.

THEOREM 60. Let R be the ring of algebraic integers in a quadratic field F, and A an ideal in R. Then $A = (a, \beta)$ where a is the generator of the ideal $A \cap Z$ of Z and $\beta \subset A$.

Proof: Check that $A \cap Z$ is an ideal in Z, and hence consists of all multiples of some integer a. Let $1, \alpha$ be an integral basis for R. Consider the set T of integers t such that $s + t\alpha \in A$ for some $s \in Z$. Then T is an ideal in Z (why?), so T consists of all multiples of some $t_0 \in Z$. Choose any $\beta \in A$ such that $\beta = s_0 + t_0\alpha$ for some $s_0 \in Z$. If $\gamma = s + t\alpha$ is any element of A, then $t \in T$ so $t = kt_0$ for some $k \in Z$. Hence, $\gamma = s + kt_0\alpha = k\beta + (s - ks_0)$, so $s - ks_0 = \gamma - k\beta \in A \cap Z$, since $\gamma, \beta \in A$. Thus, $\gamma - k\beta = na$ for some $n \in Z$, so $\gamma = na + k\beta \in (a, \beta)$.

Note that we have shown not only that $A = (a, \beta)$—that is, every element of A can be written in the form $na + k\beta$ where $n, k \in R$—but also that we can choose n and k to be *integers*. We can now show that all nonzero prime ideals are invertible.

THEOREM 61. Let R be the ring of algebraic integers in a quadratic field F, and P a nonzero prime ideal. Then P is invertible.

Proof: If $0 \neq \alpha \in P$, then $0 \neq N(\alpha) \in P \cap Z$, so $P \cap Z \neq \{0\}$. Check that $P \cap Z$ is a *prime* ideal in Z and is therefore generated by a prime number p. Thus, $P = (p, \alpha)$ for some $\alpha \in P$ by Theorem 60. Now $1 \in P^{-1}$ and $\bar{\alpha}/p \in P^{-1}$, since $\bar{\alpha}/p \cdot p = \bar{\alpha} \in R$ and $\bar{\alpha}/p \cdot \alpha = \bar{\alpha}\alpha/p \in R$ because $\bar{\alpha}\alpha \in P \cap Z$ is divisible by p. So $P^{-1}P$ contains the integers p, $\alpha + \bar{\alpha}$, and $\alpha\bar{\alpha}/p$. If $p \mid (\alpha + \bar{\alpha})$ and $p \mid \alpha\bar{\alpha}/p$, then the polynomial

$$x^2 - (\alpha + \bar{\alpha})/p \cdot x + \alpha\bar{\alpha}/p^2 = (x - \alpha/p)(x - \bar{\alpha}/p)$$

has integer coefficients and is satisfied by α/p. Thus, α/p is an algebraic integer—that is, $p \mid \alpha$ in R—so $P = (p, \alpha) = (p)$ is principal, and hence invertible. On the other hand, if $p \nmid (\alpha + \bar{\alpha})$ or $p \nmid \alpha\bar{\alpha}/p$, then, since p is prime, $P^{-1}P$ contains two relatively prime integers, so $1 \in P^{-1}P$, and $P^{-1}P = R$.

This takes care of the uniqueness of products of prime ideals in the simple case $PQ_1 = PQ_2$. Now suppose we have $P_1P_2 \cdots P_n = Q_1Q_2 \cdots Q_m$ where the P's and Q's are nonzero prime ideals. We can start canceling only if some P is equal to some Q. However, since $P_1 \supseteq P_1P_2 \cdots P_n = Q_1Q_2 \cdots Q_m$, then $P_1 \supseteq Q_i$ for some i; this follows from repeated application of property 2 of prime ideals—notice the similarity to the fact that if a prime number divides a product of integers, then it must divide one of them. We may as well relabel so that $P_1 \supseteq Q_1$. But Q_1 is a *maximal* ideal, by the corollary to Theorem 59, so $P_1 = Q_1$. Now, multiplying both sides by P_1^{-1}, we get $P_2 \cdots P_n = Q_2 \cdots Q_m$. Continuing, we match up each P with a Q, and we have shown:

THEOREM 62. Let R be the ring of algebraic integers in a quadratic field. Then a nonzero ideal in R can be written as a product of prime ideals in at most one way (aside from order of the factors).

9.4 PROBLEMS

1. Prove that if $A = (a_1, a_2, \ldots, a_n)$ is an ideal in R, then $f \in A^{-1}$ if and only if $fa_i \in R$ for $i = 1, 2, \ldots, n$.
2. Show that A^{-1} satisfies the three properties of an ideal.
3. Show that $f \in A^{-1}$ if and only if for every nonzero $a \in A$, f can be written as a quotient r/a with $r \in R$.

4. Prove that if A and B are invertible ideals, then so is AB.
5. Prove that if A is an ideal and B is a fractional ideal such that $AB = R$, then $B = A^{-1}$ (note that $B \subseteq A^{-1}$, so $A^{-1}A = R$; now look at $A^{-1}AB$).
6. Use Problem 5 to show that if A is the ideal $(3, 2 + \sqrt{-5})$ in the ring of algebraic integers in $Q[\sqrt{-5}]$, then $A^{-1} = (1, (2 - \sqrt{-5})/3)$.
7. Prove Theorem 56 for fractional ideals.
8. Let A be the ideal $(2, \sqrt{10})$ in the ring of algebraic integers in $Q[\sqrt{10}]$. Find A^{-1}.
9. Let R be a ring and S a subring of R (that is, a subset of R containing 1 that is closed under $+$, $-$, and $\cdot$). Show that if P is a prime ideal in R, then $P \cap S$ is a prime ideal in S.
10. Show that if $A \neq R$ is an invertible ideal in R, then the ideals $A, A^2, A^3, \ldots$ are all distinct.
11. Let $B \subseteq A$ be ideals in R. Show that $C = A^{-1}B$ is an ideal in R, and, if A is invertible, C is the unique solution to the (ideal) equation $AX = B$.

9.5 FACTORING INTO PRIMES

Theorem 62 says that we can factor an ideal into a product of prime ideals in *at most* one way. The question remains whether or not we can factor it into a product of prime ideals at all. If A is an ideal, where are we to look for the prime factors of A? If $A = PB$, then $P \supseteq PB = A$, so the ideals to look at are the ideals that contain A. These ideals are intimately related to the ideals in R/A.

Suppose I is an ideal containing A. Then if $r \in I$, so is $r + a$ for every element $a \in A$. Hence, if $r \equiv s \pmod{A}$, then $r \in I$ if and only if $s \in I$—that is, a residue class modulo A is either a subset of I or is disjoint from I. Denote by I/A the set of cosets of A that are subsets of I. Check that I/A is an ideal in R/A (it follows easily from the definition of the arithmetic operations in R/A and the fact that I is an ideal in R). Notice also that I is simply the union of the elements of I/A; that is, $r \in I$ if and only if r is in some element of I/A. Hence, if I_1 and I_2 are two ideals containing A such that $I_1/A = I_2/A$, then $I_1 = I_2$. The situation is depicted diagrammatically in Figure 9.1: the elements of R make up the inside of the big rectangle, the elements of A fill up the lowest slice, the elements of I constitute the shaded region. The slices are the cosets of A, the elements of R/A. The elements of I/A are the shaded slices. The unshaded slices are the cosets of A that are disjoint from I.

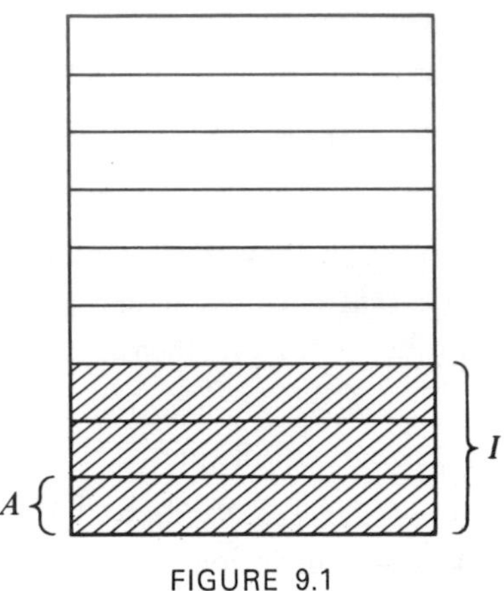

FIGURE 9.1

Conversely, suppose J is an ideal in R/A. Then J is a set of cosets of A. Let I be the union of these cosets. Check that I is an ideal in R containing A, and $J = I/A$; so every ideal in R/A is of the form I/A for some ideal I of R containing A.

THEOREM 63. Let R be a ring and A an ideal in R. If I is an ideal in R containing A, then I/A is an ideal in R/A. Moreover, every ideal in R/A has this form, and $I_1 \subseteq I_2$ if and only if $I_1/A \subseteq I_2/A$.

Proof: Only the last statement is new, and it says that $I_1 \subseteq I_2$ if and only if every coset of A contained in I_1 is contained in I_2. This is clear since I is the union of the cosets of A contained in I.

COROLLARY. If R is the ring of algebraic integers in a quadratic field, and $A \neq \{0\}$ is an ideal in R, then there are only finitely many ideals I of R containing A. If $A \neq R$, then A is contained in a prime ideal.

Proof: By Theorem 59, R/A is finite and so has only finitely many subsets, and hence only finitely many ideals. The first statement is then immediate from Theorem 63. Suppose $A \neq R$. If A is not maximal, then there is an ideal $M_1 \supseteq A$ such that $M_1 \neq A$ and $M_1 \neq R$. If M_1 is not maximal, then there is an ideal $M_2 \supseteq M_1$ (and hence $M_2 \supseteq A$) such that $M_2 \neq M_1$ and $M_2 \neq R$. If M_2 is not maximal, then there is an ideal $M_3, \ldots$, etc. We get a chain of distinct ideals $M_1, M_2, \ldots$ which has to stop because there are only finitely many ideals containing A. Hence, somewhere along the line, we must run into an ideal $M \supseteq A$ which is maximal, and therefore prime.

So if we are looking for prime factors of an ideal A, there are not too many ideals to consider, and there are always some prime ones there. This allows us to wrap up the whole deal.

THEOREM 64. If R is the ring of algebraic integers in a quadratic field and A is an ideal in R, $A \neq \{0\}$, $A \neq R$, then A is a product of (finitely many) prime ideals.

Proof: We shall show that if A is not a product of prime ideals, then there is an ideal $B \supseteq A$ such that $B \neq A$, $B \neq R$, and B is not a product of prime ideals. By repeated application, we would then get an infinite chain of distinct ideals $A \subseteq B \subseteq \ldots$ which is impossible since there are only finitely many ideals containing A. This is an induction-type argument which, when applied to Z, is known as "infinite descent." Suppose you want to show that some property is true for all positive integers. If you can show that whenever it fails for some positive integer n, then it also fails for some positive integer $m < n$, you have shown that it cannot fail, lest there be an infinite decreasing sequence of positive integers $n > m > \ldots$, which cannot be because there are only finitely many positive integers less than n. For ideals, this becomes "infinite ascent" because, roughly, the smaller integers correspond to the larger ideals.

On with the proof. By the corollary, $A \subseteq P$ where P is a prime ideal. Let $B = P^{-1}A$. Since $A \subseteq P$, we have $B \subseteq R$ and B is an ideal. If $B = R$, then, multiplying both sides by P, we get $A = P$, and we are through. If $B = A$, then, multiplying both sides by P, we get $A = PA$, and hence $P^2A = PA = A$ and, in general, $A = P^nA$ for all positive integers n. But $P^nA \subseteq P^n$, so we get $A \subseteq P^n$ for all positive n, which is impossible since the ideals P^n, being products of prime ideals, are all distinct (by Theorem 62). However, A is contained in only finitely many ideals. Thus, $B \neq R$ and $B \neq A$. Moreover, $B \supseteq A$ since $P^{-1} \supseteq R$. It remains to show that B is not a product of prime ideals, but if it were, then A would be also, because $A = PB$.

COROLLARY 1. With R as above, every nonzero ideal of R is invertible.

Proof: Every such ideal is a product of invertible ideals (see Problem 4 of the last section).

There are two notions of divisibility for ideals, depending on how we generalize the situation for principal ideals. If $A = (a)$ and $B = (b)$ are principal ideals, then $a \mid b$ if and only if $A \supseteq B$, so we might say that an ideal A divides an ideal B if $A \supseteq B$. Alternatively, since to say $a \mid b$ means $b = ac$ for some number c, this can be translated to ideals by saying that A divides B if $B = AC$ for some ideal C. In general, these need not be the same notion (see Problem 3). However, for the rings of algebraic integers in quadratic fields, they are the same.

COROLLARY 2. Let R be as above. If A and B are ideals in R, then $A \supseteq B$ if and only if $B = AC$ for some ideal C.

Proof: If $B = AC$, then certainly $A \supseteq AC = B$. Conversely, if $A \supseteq B$, then $C = A^{-1}B$ is an ideal and $B = AC$ (the case $A = \{0\}$ is trivial).

When we were looking for the right definition of prime ideals, we considered three possible properties. We chose property 2 for the definition of prime ideal and called an ideal satisfying property 3 a maximal ideal. For the algebraic integers in quadratic fields, we have seen that every nonzero prime ideal is maximal, and for any ring every maximal ideal is prime. So properties 2 and 3 are more or less the same in this situation. We can now show that property 1 also characterizes prime ideals (for quadratic fields).

COROLLARY 3. Let R be as above. Then an ideal $P \neq R$ has the property that whenever $P = AB$, then $A = R$ or $B = R$, if and only if P is a nonzero prime ideal.

Proof: If P is not a prime ideal, then P is a product of two or more prime ideals, so P can be written as AB where $A \neq R$ and $B \neq R$. Conversely, if $P = AB$ where $A \neq R$ and $B \neq R$ then, factoring A and B into prime ideals shows that P cannot be prime because of Theorem 62 (what about $\{0\}$?).

9.5 PROBLEMS

1. Let R be any commutative ring. Prove that if $A \subseteq I$ are ideals in R, then I/A is an ideal in R/A. Show that every ideal in R/A is uniquely of the form I/A for some ideal $I \supseteq A$.
2. Let R be a commutative ring and A a proper ideal of R. Show that an ideal $P \supseteq A$ is prime if and only if P/A is a prime ideal in R/A. Do the same with "prime" replaced by "maximal."
3. Let R be the ring of polynomials in one variable x with integer coefficients. Let A be the ideal of polynomials with even constant term and B the ideal of polynomials with zero constant term. Show that $A \supseteq B$ but there is no ideal C such that $B = AC$ (note that since $2 \in A$ and $B \subseteq AC$, every polynomial in C would have to have zero constant term, and hence every polynomial in AC would have to have x with an even coefficient, but $x \in B$ and 1 is not even).
4. Let p be a prime number (in Z) and R the ring of algebraic integers in a quadratic field. Then $R/(p)$ has p^2 elements by

Theorem 59. Conclude that any ideal P such that $(p) \subseteq P$ but $P \neq (p)$ and $P \neq R$ is maximal (since $R/(p)$ is an additive group with p^2 elements, any subgroup—in particular any ideal—different from 0 or the whole works has p elements).

5. With the same setup as Problem 4, show that (p) is either a prime ideal or a product of two prime ideals (if (p) is a product of three or more prime ideals, then there is a non-prime ideal satisfying the conditions in Problem 4).

6. Use the same setup as Problems 4 and 5. Show that (p) is a prime ideal in R if and only if for every $\alpha \in R$, $p \mid N(\alpha)$ implies $p \mid \alpha$ ("only if" is immediate; to prove "if," it suffices to show that if $p \mid N(\alpha)$, then $(p, \alpha) \neq R$. But if $1 = rp + s\alpha$, then $\bar{\alpha} = rp\bar{\alpha} + sN(\alpha)$, so $p \mid \bar{\alpha}$; hence $p \mid \alpha$).

7. Let R be the ring of algebraic integers in $Q[\sqrt{d}]$ where $d \not\equiv 1 \pmod{4}$ is a square-free integer. Let p be a prime number (in Z). Show that (p) is a prime ideal in R if and only if d is a quadratic nonresidue modulo p (by Problem 6, (p) is *not* prime exactly when the equation $x^2 - dy^2 \equiv 0 \pmod{p}$ has a solution in Z such that $p \nmid x$ or $p \nmid y$).

8. Let R be the ring of algebraic integers in $Q[\sqrt{-5}]$. Write the ideals (2), (5), (7), and (11) as products of prime ideals.

9. Recall (Problem 5 of section 9.1) that a ring S is a field if and only if the only ideals in S are S and $\{0\}$. Use this and Theorem 63 to prove Theorem 58: an ideal I in a ring R is maximal if and only if R/I is a field.

10. Show that every ideal in Z_n is principal.

CHAPTER X

RULER AND COMPASS CONSTRUCTIONS

10.1 EUCLIDEAN CONSTRUCTIONS

The basic objects of plane Euclidean geometry are points, lines, and circles. If we have a collection of points, lines, and circles, we may construct other points, lines, and circles by the classical Euclidean constructions:

(1) Given two distinct points, we may construct the line passing through them.

(2) Given a point A and a distance d, we may construct the circle with center A and radius d.

(3) Given two distinct lines, or two distinct circles, or a circle and a line, we may "construct" the points, if any, where they intersect.

Construction 1 is self-explanatory; it's what you can do with a *straight edge* (an unmarked ruler). Construction 2 requires a little elaboration, since it involves the notion of a distance. We are given a distance d if we are given two points B and C which are a distance d apart. This allows us to open a *compass* the right amount by placing one point on B and the other on C; we can then draw the circle whose center is A and radius is d. Construction 3 is not usually considered to be a construction since one does not have to put pencil to paper to make it; the points are already there. However, it is the only way by which we get new points to use in constructions 1 and 2. The construction consists in recognizing that the locations of these points are now specified. *The only ways to construct lines, circles, and points are by constructions* 1, 2, *and* 3, *respectively*. These are the rules for playing the ancient game of ruler and compass constructions.

Consider the following typical construction problem: given an angle α, construct the line L that bisects it. To be given an angle α means to be given a pair of intersecting lines. Actually, we would have to know somehow which of the four angles so formed is α, but we shall rely on spatial intuition and a well-placed finger for that. If we look back at our constructions, we note that with only this information the best we can do is "construct" the point of intersection of the two lines, and quit. To avoid this problem, we assume that *to be given a line means to be given at least two distinct points on it*. Similarly, *to be given a circle means to be given its center and radius*. Notice that this is true for circles and lines that arise by construction. It should be emphasized that to be given a line (or a circle) does *not* mean that we are given every point on that line. The only points available are those that are explicitly given, or are "constructed" by construction 3.

There is no real loss in generality in this assumption, because if we are given at least two distinct points anywhere, then we can construct two distinct points on any given line, and the center and radius of any given circle; and to get anywhere at all with these constructions, we must be given at least two distinct points (or be able to "construct" them immediately by construction 3). In this manner, the basic objects of geometry may be reduced to just the points; lines and circles are

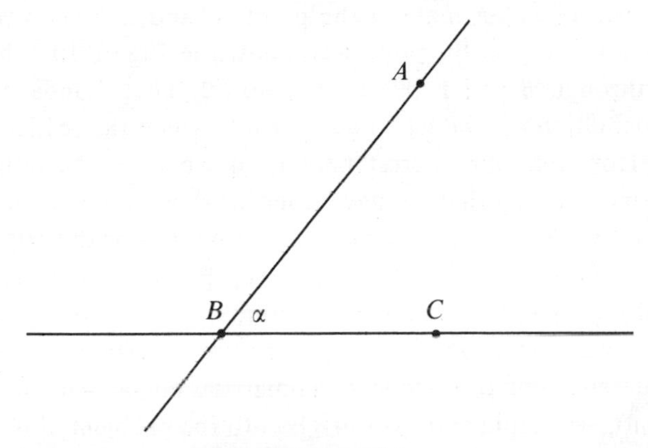

FIGURE 10.1

determined from points by constructions 1 and 2, and serve to construct new points by construction 3.

Returning to the problem of bisecting an angle, we are given three points A, B, and C, and the problem is to bisect the angle α formed by the line through A and B and the line through B and C, as in Figure 10.1.

One way to do this is as follows (see Figure 10.2): by construction 2, draw the circle with center at B and radius d, the distance from B to C. By construction 3, we get the point D on the line through A and B. Now

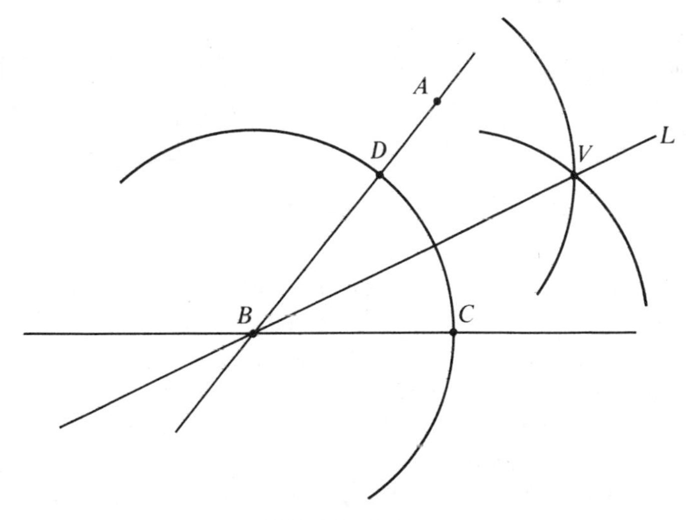

FIGURE 10.2

draw the circles of radius d about the points C and D by construction 2. By construction 3, we get the point V as shown in Figure 10.2. Now draw the line L through B and V by construction 2. That L indeed bisects α requires a definition of "bisect" and a proof, which the reader is invited to supply. However, our interest here is in what can be done, not in formally demonstrating that we have done what we wanted to.

What we have done is constructed a point V with certain desired properties, given the points A, B, and C. Any Euclidean construction can be regarded in this way: given a set of points, you construct some other points. So if we wish to know whether or not a certain construction is possible (a large chunk of Euclidean geometry is concerned with precisely this question), we should address ourselves to the problem of determining what points can be constructed from a given set of points.

Suppose we are given a collection of points including two distinct points A and B. This is enough to set up a coordinate system as follows: draw the line through A and B; this will be the x-axis. If we let the point A be the origin, and the distance between A and B be the unit of length, then the coordinates of A and B are $(0, 0)$ and $(1, 0)$. Now we can construct all the points of the form $(n, 0)$, for n an integer, using a compass as indicated in Figure 10.3.

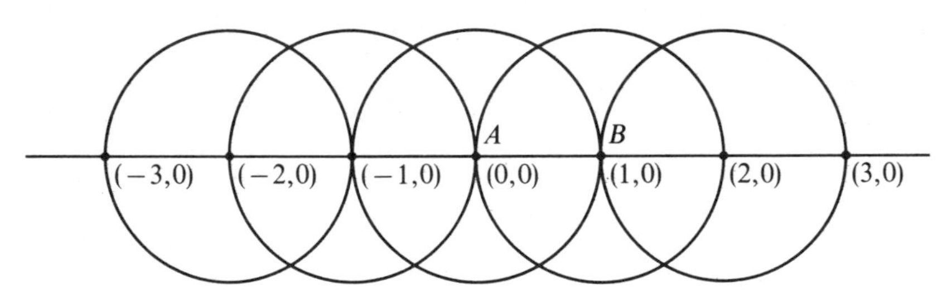

FIGURE 10.3

Next we construct the y-axis by constructing the line perpendicular to the x-axis through the point $(0, 0)$ as indicated in Figure 10.4.

We construct the points $(0, m)$, m an integer, on the y-axis just like we did on the x-axis. Once we have a coordinate system in the plane, we may use complex numbers to label the points, letting the complex number $a + bi$ correspond to the point whose coordinates are (a, b). When this is done, we frequently ignore the distinction between a point and its corresponding complex number and refer, for example, to the point $2 - 3i$ rather than to the point with coordinates $(2, -3)$. The choice

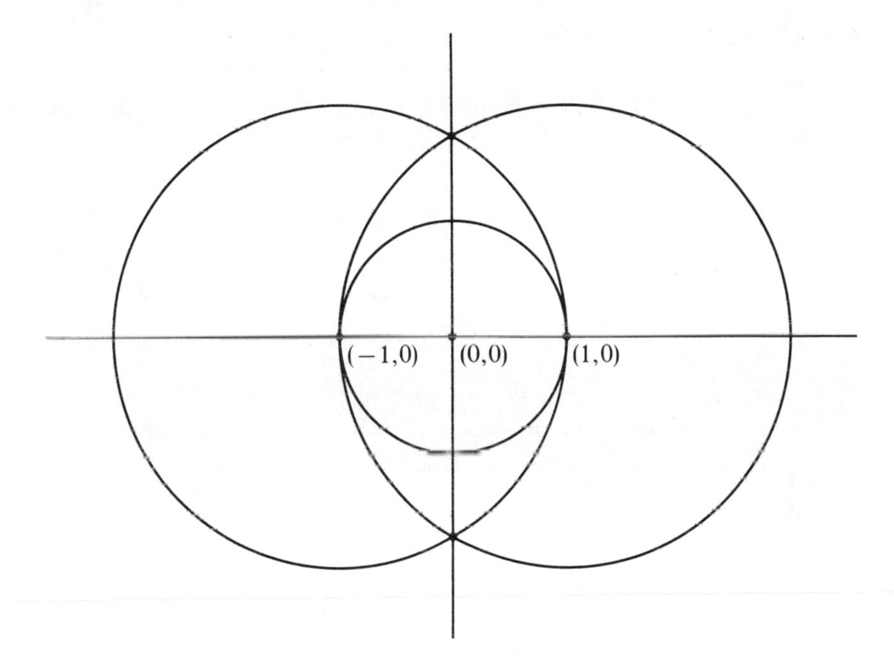

FIGURE 10.4

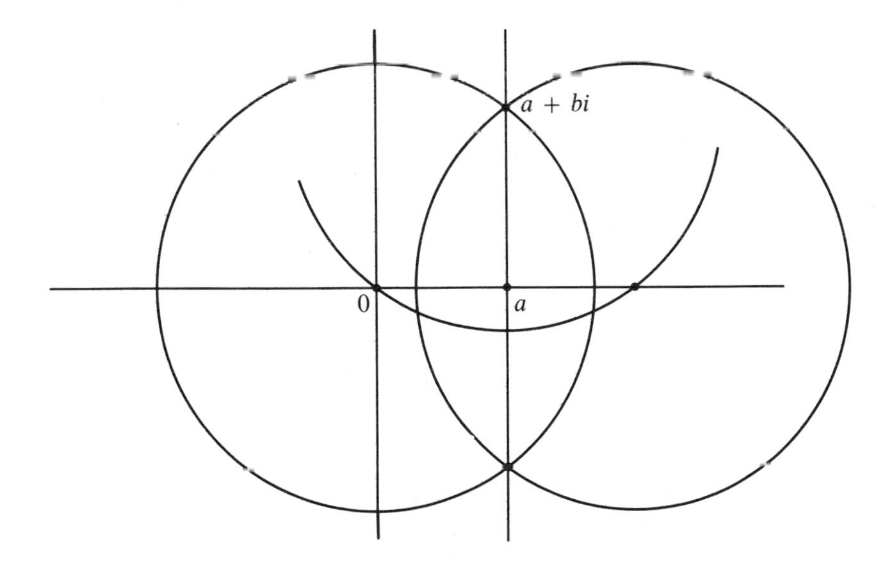

FIGURE 10.5

of the points A and B amounts to choosing which point shall be called 0 and which point shall be called 1.

THEOREM 65. The point $a + bi$ can be constructed if and only if the points a and b can be constructed.

Proof. If we have the point $a + bi$, then we can drop a perpendicular from $a + bi$ to the x-axis as indicated in Figure 10.5, where all the circles have radius equal to $\sqrt{a^2 + b^2}$, the distance from $a + bi$ to 0.
This constructs a. Similarly, by dropping a perpendicular from $a + bi$ to the y-axis, we can construct bi. Conversely, if we have a and bi, we can construct $a + bi$ as an intersection of the circle around a with radius $|b|$ (the distance from bi to 0) with the circle around bi with radius $|a|$ (the distance from a to 0). See Figure 10.6. Finally, we observe that bi is constructible if and only if b is constructible (Figure 10.7).

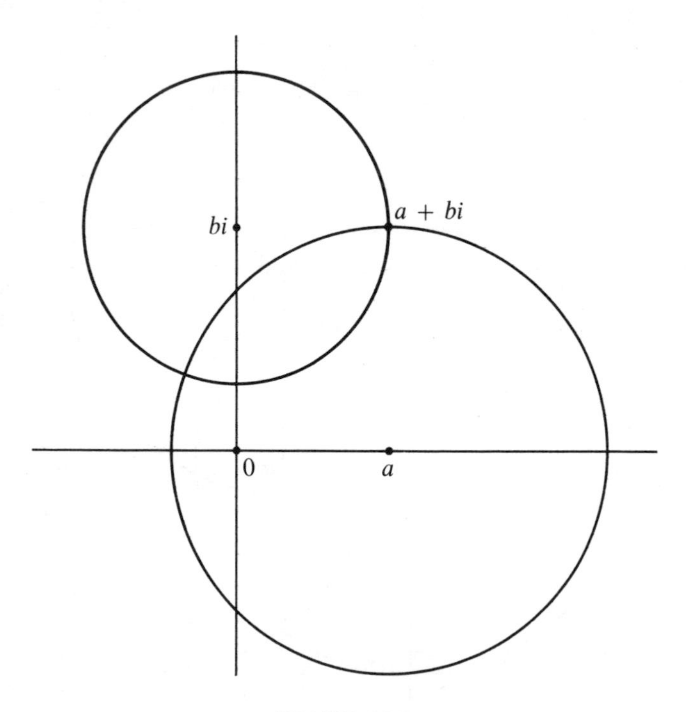

FIGURE 10.6

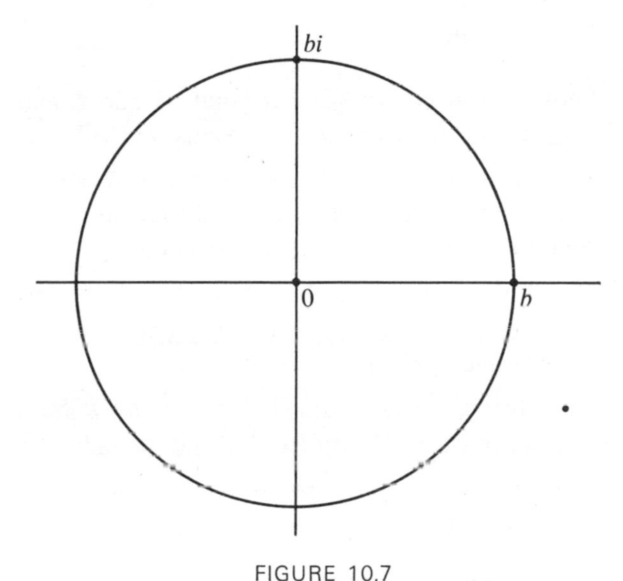

FIGURE 10.7

Where are we? Given a set S of points in the plane, we choose two of them and assign them the (complex) numbers 0 and 1. By constructing a coordinate system, we may think of the set of points S as a set of complex numbers, including the numbers 0 and 1. The points that we can construct, given S, may also be thought of as complex numbers. We noted that, given the numbers 0 and 1, we can construct all numbers of the form n and mi, where n and m are integers. By Theorem 65, we can also construct all Gaussian integers. In general, Theorem 65 allows us to concentrate on determining what *real* numbers can be constructed.

We have begun the process of translating the geometric problem of finding out what Euclidean constructions are possible to an algebraic problem about complex numbers. The idea is to interpret the geometric notions of the three basic constructions in terms of arithmetic in the complex numbers. Then we might be able to use our knowledge of algebraic numbers to figure out what can and what cannot be constructed. It is reasonable to expect algebraic numbers to play a role since both lines and circles are determined by algebraic equations ($ax + by = c$ and $(x - a)^2 + (y - b)^2 = r^2$)—that is, equations involving only arithmetic operations.

10.1 PROBLEMS

1. Show that given two distinct points A and B and a line L, you can construct two distinct points on the line L.
2. Show that given three distinct points A, B, and C, you can construct a line through A perpendicular to the line through B and C. Use this to prove Theorem 65.
3. Show that given the numbers 0 and 1, you can construct any Gaussian number (show that you can construct any rational number and apply Theorem 65).
4. Show that given two distinct points A and B and a circle C, you can construct the center of C and the radius of C.
5. Show that given the numbers 0 and 1, you can construct $\sqrt{2}$; $\sqrt{5}$; $\sqrt{3}$.
6. Show that given the numbers 0, 1, and $\sqrt{n}$, where $n > 0$, you can construct $\sqrt{n+1}$. Conclude that given the numbers 0 and 1, you can construct $\sqrt{m}$ for any positive integer m.

10.2 THE FIELD OF CONSTRUCTIBLE NUMBERS

Let S be a set of complex numbers including 0 and 1. Such a set arises from the "given" part of a geometric construction problem. We are interested in the set $\bar{S}$ of complex numbers that can be constructed from S. This set has some nice algebraic properties.

THEOREM 66. Let S be a set of complex numbers including 0 and 1. Let $\bar{S}$ be the set of complex numbers that can be constructed from S. Then $\bar{S}$ is a field, and if $\alpha^2 \in \bar{S}$ for some complex number α, then $\alpha \in \bar{S}$.

Proof: Theorem 65 says that $a + bi \in \bar{S}$ if and only if $a, b \in \bar{S}$. Moreover, the arithmetic operations on complex numbers are achieved by arithmetic operations on their (real) coordinates. So to show that $\bar{S}$ is a field, we need only show that the real numbers in $\bar{S}$ form a field. Suppose a and b are real numbers in $\bar{S}$; we shall show that $a + b$ and $a - b$ are in $\bar{S}$. If $b = 0$, then $a + b = a - b = a \in \bar{S}$, and we are through. If $b \neq 0$, place one point of the compass on 0 and the other on b. Then draw the circle with this radius around the point a; the two intersections with the x-axis are the points $a - b$ and $a + b$ as illustrated in Figure 10.8. Observe that for $a = 0$, this says that if $b \in \bar{S}$, then $-b \in \bar{S}$.

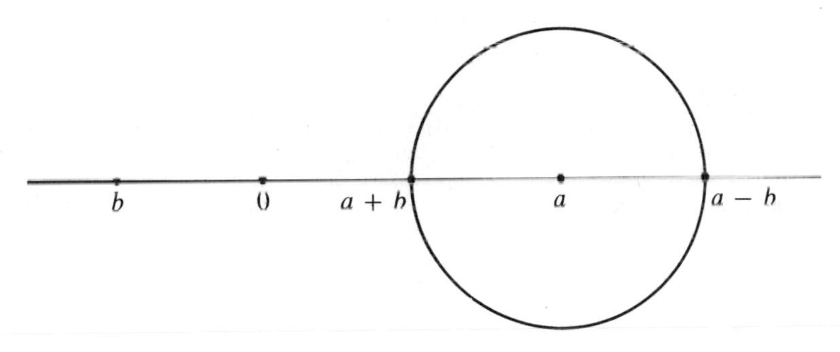

FIGURE 10.8

Next, we show that if a and b are positive real elements of $\bar{S}$, then $a/b \in \bar{S}$. By Theorem 65, we can construct the point $b + i$. Draw the line L through 0 and $b + i$. Now construct the line K through a, perpendicular to the x-axis (see Figure 10.9). It is an

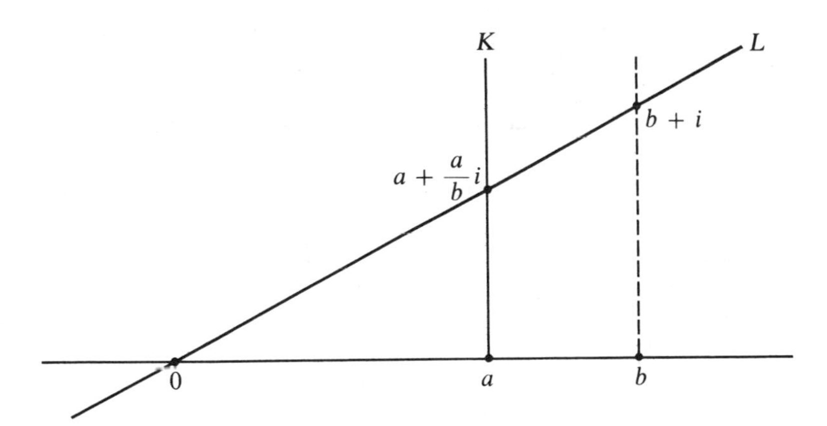

FIGURE 10.9

elementary exercise in similar triangles to deduce that the intersection of L and K is the point $a + (a/b)i$, whether $a < b$ or $b \leq a$. Hence, by Theorem 65, $a/b \in \bar{S}$.

From the first part of the proof, a real number is in $\bar{S}$ if and only if its negative is; hence, for any real numbers a and b in $\bar{S}$, if $b \neq 0$, then $a/b \in \bar{S}$. Since $a0 = 0$ and, if $b \neq 0$, $ab = a/(1/b)$, then it also follows that if a and b are real numbers in $\bar{S}$, then $ab \in \bar{S}$. Thus, the real numbers in $\bar{S}$ form a field, and hence $\bar{S}$ is a field.

Finally, suppose $\alpha^2 = a + bi \in \bar{S}$. Since the polynomial $x^2 - (a + bi)$ has at most two complex roots, and you can readily verify that

$$\pm\left(\sqrt{\frac{1}{2}(a + \sqrt{a^2 + b^2})} + i\sqrt{\frac{1}{2}(-a + \sqrt{a^2 + b^2})}\right)$$

are roots, then

$$\alpha = \pm\left(\sqrt{\frac{1}{2}(a + \sqrt{a^2 + b^2})} + i\sqrt{\frac{1}{2}(-a + \sqrt{a^2 + b^2})}\right).$$

By Theorem 65, both a and b are in $\bar{S}$, so, in order to show $\alpha \in \bar{S}$, it suffices to show that $\sqrt{r} \in \bar{S}$ whenever r is a positive real number in $\bar{S}$. To do this, draw the circle of radius $(1 + r)/2$ around the point $(1 + r)/2$, and the line perpendicular to the x-axis through the point 1 (see Figure 10.10). Let $1 + yi$ be an intersection. Then

$$\left(1 - \frac{1}{2}(1 + r)\right)^2 + y^2 = \left(\frac{1}{2}(1 + r)\right)^2$$

so $y^2 = r$, and hence $y = \pm\sqrt{r} \in \bar{S}$.

Theorem 66 says that the set $\bar{S}$ of complex numbers that can be constructed from S is closed under addition, subtraction, multiplication, division (by nonzero numbers), and taking square roots. In light of Theorem 65, and the proof of Theorem 66, the same statement can be made about coordinates. In particular, since $S \subseteq \bar{S}$, *any point whose coordinates can be expressed in terms of the coordinates of the points in S and the operations $+$, $-$, $\cdot$, $\div$, and $\sqrt{\ }$, can be constructed from S.* For example, given the points 0 and 1 we can construct the point

$$\frac{\sqrt{\sqrt{3} - \sqrt{1} + \sqrt{2}}}{1/4 + \sqrt{1/2 + \sqrt{5}}} + \frac{1/2\sqrt{17} - \sqrt{\sqrt{2}}}{\sqrt{\sqrt{11}} - 1/2\sqrt{13}}\, i.$$

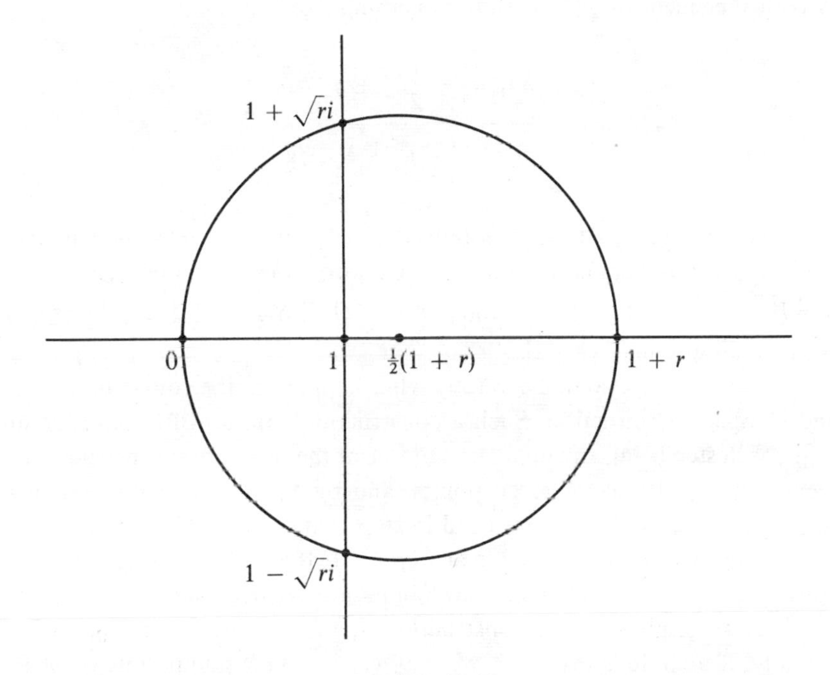

FIGURE 10.10

The striking fact is that no other points can be constructed from S. To see this, we watch closely how elements of $\bar{S}$ are built up from elements of S. The simplest elements of $\bar{S}$ are those whose coordinates can be expressed in terms of the coordinates of S and the arithmetic operations alone, with no taking of square roots. Although it is not strictly necessary, we shall assume that S is a finite set. Notice that in any given construction, only finitely many points of S are used, so even if S were infinite, we could get by with looking at only a finite subset of its elements.

DEFINITION. If $S = \{a_1 + b_1 i, a_2 + b_2 i, \ldots, a_n + b_n i\}$ is a set of complex numbers, then the *coordinate field of* S is defined to be the set of all quotients c/d, where $c, d \in Q[a_1, \ldots, a_n, b_1, \ldots, b_n]$ and $d \neq 0$. This field is denoted by $Q(S)$.

Recall that $Q[a_1, \ldots, a_n, b_1, \ldots, b_n]$ is the set of all numbers which can be written as polynomials in $a_1, \ldots, a_n, b_1, \ldots, b_n$ with coefficients in Q.

A typical element of $Q(S)$ is thus something like

$$\frac{5a_1{}^3a_2 - \frac{1}{2}b_3 + a_2{}^2b_4}{7b_1 + \frac{1}{4}a_1b_2{}^2b_4{}^3 - 18}$$

You should check that $Q(S)$ is indeed a field; in fact, it is the smallest field of real numbers containing all the coordinates of the points in S. If $S = \{0, 1\}$ or $S = \{0, 1, 2 + i\}$, then $Q(S) = Q$; if $S = \{0, 1, \sqrt{2}\}$, then $Q(S)$ is the quadratic field $Q[\sqrt{2}]$.

Now let's examine carefully what happens in the course of a ruler and compass construction. Such a construction consists of a sequence of steps, each step being an application of one of the three basic constructions. Denote by S_0 the set of given points, and by S_j the set of given points together with all points constructed in steps 1 through j. Thus, S_j consists of the totality of points available at step $j + 1$. If step j is an application of constructions 1 or 2, then no new points are created and $S_j = S_{j-1}$. If step j is an application of construction 3, then we get a new point s_j; actually, if a circle is involved, we may get two new points, but we shall view this as two separate applications of construction 3. If the construction ends after n steps, we get a sequence of sets $S_0 \subseteq S_1 \subseteq S_2 \subseteq \cdots \subseteq S_n$ such that either $S_j = S_{j-1}$ or $S_j = S_{j-1} \cup \{s_j\}$, where s_j is the point constructed in step j. What about the coordinate fields? Clearly $Q(S_{j-1}) \subseteq Q(S_j)$. Moreover, these fields are closely related algebraically.

THEOREM 67. With the notation of the preceding paragraph, either $Q(S_j) = Q(S_{j-1})$ or $Q(S_j) = Q(S_{j-1})[\sqrt{v}]$ for some $v \in Q(S_{j-1})$.

Proof: If $S_j = S_{j-1}$, then $Q(S_j) = Q(S_{j-1})$. Suppose $S_j = S_{j-1} \cup \{s_j\}$, where s_j is the point constructed at step j. Then s_j is constructed as the intersection of two lines, or the intersection of a line and a circle, or the intersection of two circles. These lines and circles must have been constructed before step j. Any such line must pass through two distinct points in S_{j-1}, and hence has an equation $ax + by + c = 0$, where $a, b, c \in Q(S_{j-1})$. Any such circle must have its center in S_{j-1} and radius equal to the distance between two points in S_{j-1}. Hence, such a circle has an equation $(x-d)^2+(y-e)^2=r^2$ where $d, e, r^2 \in Q(S_{j-1})$. If $s_j = x + yi$ lies on two such lines, then its coordinates satisfy both equations and, upon solving these two equations, it is readily seen that the coordinates of s_j are already in $Q(S_{j-1})$, and so $Q(S_j) = Q(S_{j-1})$. If s_j lies on such a line and such a circle, then its coordinates satisfy both equations. Eliminating one variable, say y, with the equation of the line, the equation of the circle becomes a quadratic equation in x with coefficients in $Q(S_{j-1})$. The solution will be of the form $x = (-u \pm \sqrt{v})/2w$, where $u, v, w \in Q(S_{j-1})$, $w \neq 0$. Using the equation of the line to solve for y gives $y = u_0 + w_0\sqrt{v}$ where $u_0, w_0 \in Q(S_{j-1})$. Hence, $Q(S_j) \subseteq Q(S_{j-1})[\sqrt{v}]$. But $\sqrt{v} = \pm(2wx + u) \in Q(S_j)$, so $Q(S_j) = Q(S_{j-1})[\sqrt{v}]$. The two-circle case reduces immediately to the line-circle case upon subtracting the equations of the two circles.

This theorem, together with Theorem 66, yields an algebraic characterization of which Euclidean constructions are possible.

COROLLARY 1. Let S be a set of complex numbers including 0 and 1. Then a complex number $a + bi$ can be constructed from S if and only if there exist fields $Q(S) = F_0 \subseteq F_1 \subseteq \cdots \subseteq F_n$ such that $a, b \in F_n$ and $\dim_{F_{j-1}} F_j = 2$, for $j = 1, 2, \ldots, n$.

Proof: Suppose such fields $F_0, F_1, \ldots, F_n$ exist. By Theorems 65 and 66, the elements of $F_0 = Q(S)$ can be constructed from S and, by Theorem 65, $a + bi$ can be constructed from $a, b \in F_n$. We need only show that the elements of F_j can be constructed from F_{j-1}. But, since $\dim_{F_{j-1}} F_j = 2$, if $\alpha \in F_j$ then α is algebraic over F_{j-1} of degree at most 2, by Corollary 3 to Theorem 43. Solving the quadratic equation given by setting the minimal polynomial of α over F_{j-1} equal to zero shows that $\alpha = c + d\sqrt{v}$, where $c, d, v \in F_{j-1}$. By Theorem 66, α is constructible from $\{0, 1, c, d, v\} \subseteq F_{j-1}$.

Conversely, suppose $a + bi$ can be constructed from S. By Theorem 65, a and b can be constructed from S. So there is a sequence of sets $S = S_0 \subseteq S_1 \subseteq \cdots \subseteq S_n$ as in Theorem 67 such that $a, b \in {}_nS$. Let $F_j = Q(S_j)$. By Theorem 67, $F_j = F_{j-1}$ or $F_j = F_{j-1}[\sqrt{v}]$ for some $v \in F_{j-1}$. Since $\sqrt{v}$ is algebraic over F_{j-1} of degree at most 2, then either $F_j = F_{j-1}$ or $\dim_{F_{j-1}} F_j = 2$. Upon throwing out the fields F that occur more than once, and reindexing, we get the desired sequence of fields.

Although this characterization may seem a bit awkward, it provides the following easily applied criterion for showing that a point cannot be constructed.

COROLLARY 2. Let S be a set of complex numbers including 0 and 1. If $a + bi$ can be constructed from S, then a and b are algebraic over $Q(S)$ of degree a power of 2.

Proof: Repeated application of Theorem 45 shows that $\dim_{Q(S)} F_n$ is a power of 2; in particular, F_n is finite-dimensional over $Q(S)$, so, since $a, b \in F_n$, we may apply Corollary 3 to Theorem 43 and conclude that a and b are algebraic over $Q(S)$. Letting $K = Q(S)[a]$ or $Q(S)[b]$ we have, by Theorem 45 again, $2^m = \dim_{Q(S)} F_n = (\dim_{Q(S)} K)(\dim_K F_n)$, so $\dim_{Q(S)} K$ is a power of 2. Hence, by Corollary 1 to Theorem 45 (where $F = Q(S)$ and $m = 1$), a and b are algebraic over $Q(S)$ of degree a power of 2.

10.2 PROBLEMS

1. Let $\bar{S}$ be a set of complex numbers such that $a + bi \in \bar{S}$ if and only if $a, b \in \bar{S}$. Show that $\bar{S}$ is a field if and only if the real numbers in $\bar{S}$ form a field.
2. Derive the formula for α in the proof of Theorem 66 by setting $\alpha = c + di$ and solving the equations given by $\alpha^2 = a + bi$ for c and d.
3. Write out, step by step, the construction of the bisector of an angle, given in section 10.1, when $A = 1 + i$, $B = 0$, and $C = 1$, keeping track of the field $Q(S_j)$ at each step. Verify that $Q(S_j) = Q(S_{j-1})$ or $Q(S_j) = Q(S_{j-1})[\sqrt{v}]$ for some $v \in Q(S_{j-1})$.

4. Let S be a set of complex numbers including 0 and 1. Show that any line passing through two distinct points of S has an equation of the form $ax + by + c = 0$, where $a, b, c \in Q(S)$. Show that any line with such an equation can be constructed from S.

5. Let S be a set of complex numbers including 0 and 1. Show that any circle whose center is in S and whose radius is the distance between two points in S has an equation of the form $(x - a)^2 + (y - b)^2 = c$, where $a, b, c \in Q(S)$. Show that any circle with such an equation can be constructed from S.

6. In the proof of Theorem 67, write out the details that show that $Q(S_{j-1}) = Q(S_j)$ if s_j is constructed by intersecting two lines.

10.3 THE CLASSICAL CONSTRUCTION PROBLEMS

The most famous construction problems are those of "trisecting an angle," "squaring a circle," and "duplicating a cube."

(1) Given an angle, construct the two lines that cut it into three equal angles.
(2) Given a circle, construct a square of the same area.
(3) Given a cube, construct a cube of twice the volume.

None of these constructions is possible by the Euclidean methods we have restricted ourselves to. The easiest to eliminate is duplication of a cube. By "given a cube," we mean we are given one of its edges—that is, two points whose distance apart is the length of an edge of the cube. We are required to construct two points whose distance apart is the length of an edge of a cube of twice the volume. Coordinatizing this problem, the edge of the given cube is the line segment from 0 to 1, so we are given the points 0 and 1. The given cube then has a volume of 1, measured in terms of our chosen unit of length. We want to construct a cube whose volume is 2, which requires constructing two points α and β such that the distance from α to β is $\sqrt[3]{2}$, the length of an edge of a cube of volume 2.

If we could do this, we could then place one point of a compass on α and one on β and then draw a circle of radius $\sqrt[3]{2}$ around 0, constructing the *point* $\sqrt[3]{2}$ as an intersection of this circle with the x-axis. But the minimal polynomial of $\sqrt[3]{2}$ over Q, the coordinate field of the given points, is $x^3 - 2$ (why?). Thus, $\sqrt[3]{2}$ is of degree 3 over Q, and hence cannot be constructed from 0 and 1, by Corollary 2 to Theorem 67.

We have to be a little more careful with what we mean by "you can't trisect an angle." In fact you *can* trisect certain angles, like a 90° angle (see Problem 1); and if the right points happen to be available, you could trisect any angle. What is meant is that you can be given an angle that you cannot trisect.

Suppose we are given the points 0, 1, and $1 + \sqrt{3}\,i$. We need a little trigonometry here. The angle formed at 0 is 60°. To trisect this angle, we must construct the line L passing through 0 at an angle of 20° with the x-axis. But L intersects the circle of radius 1 around 0 at the point $\cos 20° + i \sin 20°$ (see Figure 10.11).

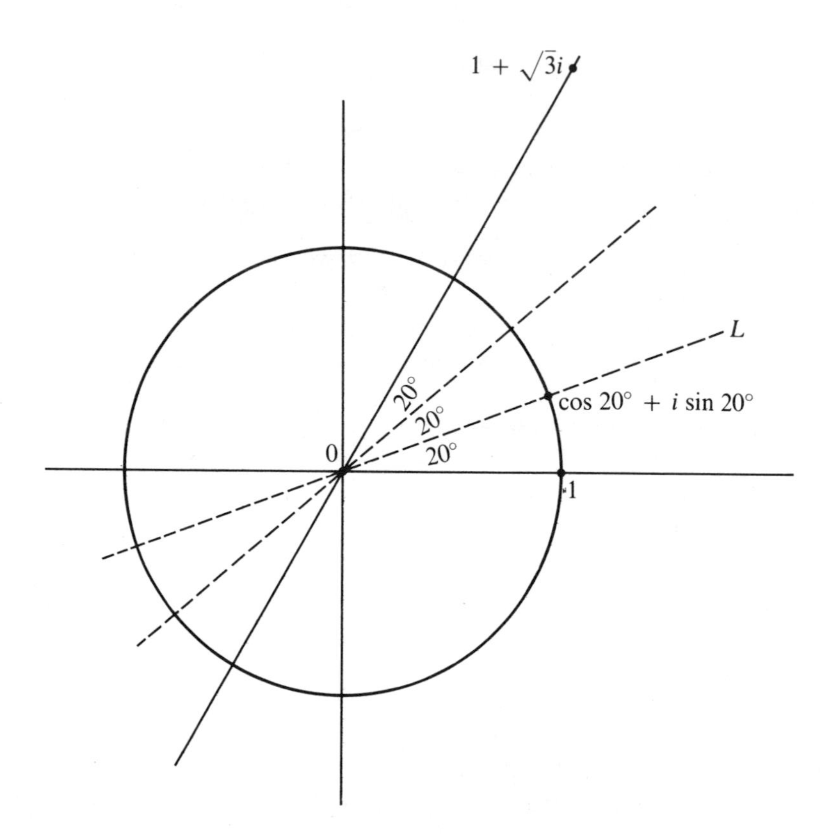

FIGURE 10.11

Hence, by Theorem 65, if we could construct L from the points 0, 1, and $1 + \sqrt{3}\, i$, we could construct the number $\cos 20°$ from the same points. But by Theorem 66, $1 + \sqrt{3}\, i$ can be constructed from 0 and 1; so if we can construct $\cos 20°$ from $\{0, 1, 1 + \sqrt{3}\, i\}$ we can already construct it from $\{0, 1\}$. Then by Corollary 2 to Theorem 67, $\cos 20°$ would have to be algebraic over Q of degree a power of 2. We shall show that that is not the case.

Since $\sin 2y = 2 \sin y \cos y$ and $\cos 2y = 2 \cos^2 y - 1$, we have

$$\cos 3y = \cos 2y \cos y - \sin 2y \sin y = 2 \cos^3 y - \cos y - 2 \sin^2 y \cos y.$$

Eliminating the $\sin^2 y$ by the identity $\sin^2 y = 1 - \cos^2 y$ we get $\cos 3y = 4 \cos^3 y - 3 \cos y$. Plugging in $y = 20°$, we get

$$1/2 = 4 \cos^3 20° - 3 \cos 20°$$
$$\text{or} \quad 8 \cos^3 20° - 6 \cos 20° - 1 = 0.$$

Dividing through by $-\cos^3 20°$, we see that $1/\cos 20°$ satisfies the polynomial $x^3 + 6x^2 - 8$. If this polynomial had a root in Q, it would have a root in Z, by Theorem 47. You only have to test $x = -6, -5, -4, -3, -2, -1, 0, 1, 2$ to prove that it has no roots in Z (why?). If $x^3 + 6x^2 - 8$ had a nontrivial factorization in $Q[x]$, it would have a root in Q (look at the degrees of the factors). Since it does not, it is irreducible, and so is the minimal polynomial of $1/\cos 20°$ over Q (the minimal polynomial must divide it). Hence, $1/\cos 20°$ has degree 3 over Q and so cannot be constructed from $\{0, 1\}$ by Corollary 2 to Theorem 67. By Theorem 66, neither can $\cos 20°$.

What about squaring a circle? Given any circle, we may choose coordinates so it is the circle of radius 1 around 0. We may as well assume we are given just the points 0 and 1. So we want to construct a square of area $\pi r^2 = \pi$. To do this, we must construct two points, α and β, such that the distance from α to β is $\sqrt{\pi}$. As in the argument against duplicating a cube, this would allow us to construct the point $\sqrt{\pi}$. By Corollary 2 to Theorem 67, we could conclude that $\sqrt{\pi}$, and hence π, was algebraic over Q of degree a power of 2. This is false; in fact, π is not even algebraic over Q. Unfortunately, any proof of this fact uses nonalgebraic techniques and would carry us too far afield. The problem is that we want to prove that there is no algebraic way of getting at π, so we could hardly hope for the proof to be algebraic. Contrast this to $\sqrt[3]{2}$ and cos 20° which are both algebraic, but of the wrong degree.

10.3 PROBLEMS

1. Show that $x^3 - 2$ is irreducible over Q. Why does this imply that $\sqrt[3]{2}$ is of degree 3 over Q?
2. Show that you cannot trisect a 30° angle.
3. Show that you cannot construct an angle of 1° from $\{0, 1\}$.

10.4 REGULAR POLYGONS

The problem of trisecting an angle is intimately related to another problem with a long history: constructing regular polygons. A regular polygon is formed by placing n equally spaced points on a circle and joining adjacent points with line segments. If we wish to specify n, we talk of a *regular n-gon.* Figure 10.12 illustrates a regular 6-gon and a regular 8-gon, usually referred to by their respective Greek names: hexagon and octagon.

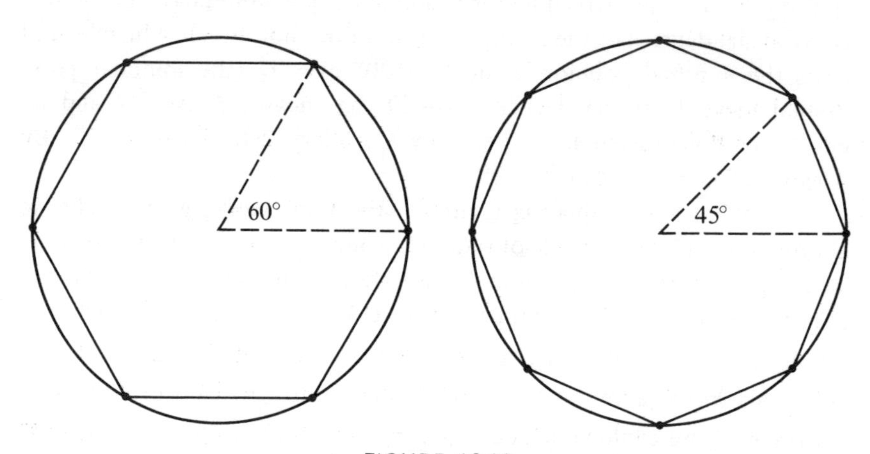

FIGURE 10.12

The problem of constructing a regular n-gon is simply the problem of constructing a $(360/n)°$ angle from scratch. Here "scratch" means given the minimal material: two points, which we may coordinatize by the complex numbers 0 and 1. Aside from the square, or regular 4-gon, two basic constructions were known to the ancient Greeks: the regular 3-gon, or triangle, and the regular 5-gon, or pentagon. Combining these constructions allows you to construct a regular 15-gon, and by bisecting angles you can pass from a regular n-gon to a regular $2n$-gon. Here the problem rested for 2000 years. Notice that the impossibility of trisecting a 60° angle precludes the construction of a 20° angle and hence the construction of a regular 18-gon.

The problem is best interpreted algebraically as that of constructing nth roots of 1. If P is a point on the *unit circle* (the circle of radius 1 around 0), then $P = \cos\theta + i\sin\theta$ where θ is the angle between the x-axis and the line through 0 and P (see Figure 10.13). For every angle θ there is a unique such point, which we indicate by $P(\theta)$. Then $P(0°) = P(360°) = 1$ and the "addition formulas" for sine and cosine say that

$$P(\theta)P(\psi) = P(\theta + \psi) \tag{1}$$

where the multiplication is just multiplication of complex numbers.

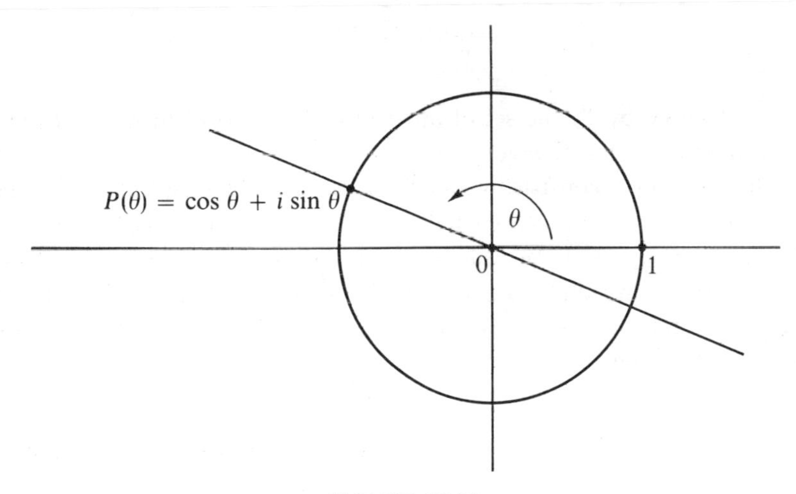

FIGURE 10.13

If $\theta = (360/n)°$, then the n points $P(\theta)$, $P(2\theta)$, $P(3\theta)$, . . . , $P(n\theta) = 1$ are equally spaced around the unit circle, and so are the vertices of a regular n-gon. Repeated application of formula 1 shows that if $\xi = P(\theta)$, then $P(j\theta) = \xi^j$, $j = 1, 2, \ldots, n$. Since $\xi^n = 1$, we have $(\xi^j)^n = (\xi^n)^j = 1^j = 1$, so these points are simply the n nth roots of 1, and ξ is a primitive nth root of 1. Figure 10.14 illustrates the case $n = 5$.

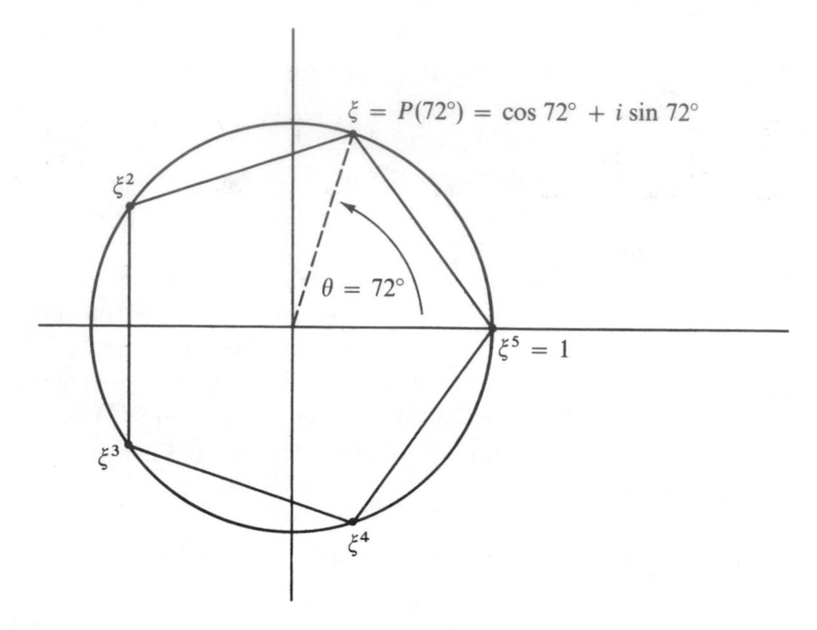

FIGURE 10.14

Denote by P_n the set of nth roots of 1. The elements of P_n are the vertices of a regular n-gon inscribed in the unit circle, and it is clear that the problem of constructing a regular n-gon is the problem of whether or not P_n is constructible. To find out whether P_n is constructible, we investigate $Q(P_n)$, and we attack $Q(P_n)$ through $Q[\xi]$. Now the fields $Q[\xi]$ and $Q(P_n)$ are different; indeed, the elements of $Q(P_n)$ are all real whereas ξ is not (unless $n = 2$). However, if we adjoin i to either field, we get the same thing.

THEOREM 68. If ξ is a primitive nth root of 1, then $(Q[\xi])[i] = Q(P_n)[i]$.

Proof: We shall show that $\xi \in Q(P_n)[i]$ and $Q(P_n) \subseteq (Q[\xi])[i]$. Let $\xi = \cos\theta + i \sin\theta$. Then since $\xi \in P_n$, its coordinates, $\cos\theta$ and $\sin\theta$, are in $Q(P_n)$, so $\xi \in Q(P_n)[i]$. On the other hand,

$$\xi(\cos\theta - i \sin\theta) = \cos^2\theta + \sin^2\theta = 1 = \xi\xi^{n-1},$$

so $\xi^{n-1} = \cos\theta - i\sin\theta \in Q[\xi]$. Hence, $\cos\theta = (\xi + \xi^{n-1})/2$ and $i\sin\theta = (\xi - \xi^{n-1})/2$ are in $Q[\xi]$. Thus, $\cos\theta$ and $\sin\theta = (-i\sin\theta)i$ are in $(Q[\xi])[i]$. Since $P_n = \{1, \xi, \ldots, \xi^{n-1}\}$ and the coordinates of any power of ξ are polynomials in $\cos\theta$ and $\sin\theta$, we have $Q(P_n) \subseteq (Q[\xi])[i]$.

COROLLARY. If P_n is constructible, then the degree of ξ over Q is a power of 2.

Proof: By Corollary 2 to Theorem 67, if P_n is constructible, then the dimension of $Q(P_n)$ over Q is a power of 2. Since i is of degree 2 over $Q(P_n)$ (why?), $Q(P_n)[i]$ is of dimension 2 over $Q(P_n)$ and hence, by Theorem 45, is of dimension a power of 2 over Q. Since $\xi \in (Q[\xi])[i] = Q(P_n)[i]$, the degree of ξ is a power of 2 by Corollary 1 to Theorem 45 (where $m = 1$).

To make use of this corollary, we need to know what the degree of a primitive nth root of 1 is. Let ξ be a primitive nth root of 1. Then the set $P_n = \{1, \xi, \xi^2, \ldots, \xi^{n-1}\}$ of nth roots of 1 is a group under multiplication (why?) and an element $\alpha \in P_n$ is primitive if and only if the order of α is n (why?). If $\alpha = \xi^j$, then α has order n if and only if $(j, n) = 1$ by Theorem 13. Thus, there are $\phi(n)$ primitive nth roots of 1.

Let $f(x)$ be the minimal polynomial of ξ. We shall show that the roots of $f(x)$ are precisely the primitive nth roots of 1, and hence the degree of $f(x)$, and therefore the degree of ξ, is $\phi(n)$. Let α be a root of $f(x)$. Since $f(x)$ is the minimal polynomial of ξ, it is irreducible and hence is also the minimal polynomial of α. Therefore, $\xi^j - 1 = 0$ if and only if $\alpha^j - 1 = 0$, since either equation is equivalent to $f(x)$ dividing $x^j - 1$. Thus, $\alpha^n = 1$ and $\alpha^m \neq 1$ if $m < n$, so α is a primitive nth root of 1.

It remains to show that *every* primitive nth root of 1 is a root of $f(x)$. Since ξ^j is primitive if and only if $(j, n) = 1$, it suffices to show that if α is a root of $f(x)$, then so is α^q for any prime q such that $q \nmid n$ (look at the prime factorization of $j = q_1 q_2 \cdots q_t$ and pass from ξ to ξ^j by the sequence $\xi, \xi^{q_1}, (\xi^{q_1})^{q_2}, \ldots, \xi^j$).

If you haven't yet done Problem 6 of section 7.2, you should do it now; we shall be using unique factorization into primes for polynomial rings over a field. By Corollary 2 to Theorem 48, we can write $x^n - 1$ as a product of irreducible monic polynomials with integer coefficients. Since

$f(x)$ is irreducible, monic, and divides $x^n - 1$, we may write this factorization as $x^n - 1 = f(x)g(x) \cdots h(x)$. If $f(\alpha) = 0$, then, since x^q is also a root of $x^n - 1$, it must be a root of one of the prime factors of $x^n - 1$. We want to show that $f(\alpha^q) = 0$. If not, then α^q is a root of some other prime factor, say $g(x)$. If $g(\alpha^q) = 0$, then α is a root of the polynomial $g(x^q)$, and so $f(x) \mid g(x^q)$. To get a contradiction out of this, we want to relate $g(x^q)$ to $g(x)$ in some simple way. The plan is to pass from polynomials with integer coefficients to polynomials with coefficients in Z_q where raising to the qth power has a lovely property.

LEMMA 1. Let q be a prime number and R a ring such that $qr = 0$ for every $r \in R$. Then if $a, b \in R$, $(a + b)^q = a^q + b^q$.

Proof: Multiplying out $(a + b)^q$, we get

$$a^q + qa^{q-1}b + \frac{1}{2}q\,(q-1)a^{q-2}b^2 + \cdots + qab^{q-1} + b^q.$$

Since q is a prime, the numbers $q, q(q-1)/2, \ldots,$ are all multiples of q. Hence, since $qr = 0$ for every r in R, all the terms in the middle are zero.

LEMMA 2. Let q be a prime number and $G(x)$ a polynomial with coefficients in Z_q. Then the polynomials $G(x^q)$ and $G(x)^q$ are equal.

Proof: Note that $Z_q[x]$ is a ring satisfying the hypothesis of Lemma 1. Let $G(x) = a_n x^n + a_{n-1}x^{n-1} + \cdots + a_1 x + a_0$. Then by repeated application of Lemma 1, we get

$$G(x)^q = a_n{}^q x^{nq} + a_{n-1}{}^q x^{(n-1)q} + \cdots + a_1{}^q x^q + a_0{}^q.$$

But by Fermat's Theorem, $a_j{}^q = a_j$, and the lemma is proved.

Let F and G be the elements of $Z_q[x]$ gotten from f and g by looking at the coefficients modulo q. Since $f(x) \mid g(x^q)$, we have $g(x^q) = f(x)g_0(x)$ and, by Theorem 48, $g_0(x)$ has integer coefficients. By looking at modulo q, we see that $G(x^q) = F(x)G_0(x)$ where $G_0 \in Z_q[x]$. Hence,

$F(x) \mid G(x^q)$, so $F(x) \mid G(x)^q$ by Lemma 2. Now F and G may not be irreducible, even though f and g were. However, we shall show that F and G have no common factors, which contradicts $F \mid G^q$ since we have unique factorization in $Z_q[x]$. If F and G had a common factor, then $x^n - 1 = F(x)G(x) \cdots H(x)$, as an element of $Z_q[x]$, would be divisible by some square. We shall show that this cannot be, thus wrapping up the proof that the degree of ξ is $\phi(n)$.

LEMMA 3. Let q be a prime number such that $q \nmid n$, and let $K(x)$ be a monic polynomial with coefficients in Z_q. If $K(x)^2$ divides $x^n - 1$ in $Z_q[x]$, then $K(x) = 1$.

Proof: Suppose $x^n - 1 = K(x)^2 L(x)$ for some $L(x) \in Z_q[x]$. Then, differentiating both sides (see Problem 7 of section 7.2), we get

$$\begin{aligned} nx^{n-1} &= 2K(x)K'(x)L(x) + K(x)^2 L'(x) \\ &= K(x)(2K'(x)L(x) + K(x)L'(x)), \end{aligned}$$

so $K(x)$ divides nx^{n-1}. Since $K(x)$ is monic and $nx^{n-1} \neq 0$ (this is where we need $q \nmid n$), we must have $K(x) = x^j$, where $0 \leq j \leq n-1$. But if $j > 0$, then x^j does not divide $x^n - 1$, whereas $x^n - 1 = K(x)^2 L(x)$ is divisible by $K(x)$. Hence, $j = 0$ and so $K(x) = 1$.

That's the end of a long road. We have proved:

THEOREM 69. If ξ is a primitive nth root of 1, then the degree of ξ is $\phi(n)$.

COROLLARY 1. If P_n is constructible, then $\phi(n)$ is a power of 2.

Proof: Immediate from the Corollary to Theorem 68, and Theorem 69.

COROLLARY 2. If P_n is constructible, then $n = 2^m p_1 p_2 \ldots p_s$ where m is a nonnegative integer and the p's are distinct primes of the form $2^t + 1$.

Proof: The problem is to determine when $\phi(n)$ is a power of 2. By Corollary 3 to Theorem 22, $\phi(n)$ is a power of 2 if and only if $\phi(p^i)$ is a power of 2 for every prime power p^i occurring in the factorization of n. But $\phi(p^i) = p^{i-1}(p-1)$, by Theorem 23, and this is a power of 2 if and only if $p = 2$, or $i = 1$ and $p - 1$ is a power of 2. Hence, $\phi(n)$ is a power of 2 if and only if n has the form described. Corollary 2 then follows from Corollary 1.

Primes of the form $2^t + 1$ are called *Fermat primes*. The first five are 3, 5, 17, 257, and 65537, corresponding to $t = 1, 2, 4, 8$, and 16. These are the only known Fermat primes.

10.4 PROBLEMS

1. Construct a regular 3-gon and a regular 4-gon.
2. Constructibility of a regular pentagon, I.
   (a) Show that P_5 is constructible if and only if $\cos 72°$ is.
   (b) Show that $\cos 36° - \cos 72° = 1/2$ (add the equations $\cos 72° = 2\cos^2 36° - 1$ and $\cos 36° = 1 - 2\sin^2 18° = 1 - 2\cos^2 72°$).
   (c) Use (b) together with $\cos 72° = 2\cos^2 36° - 1$ to show that $\cos 72°$ is a root of the polynomial $x^2 + x/2 - 1/4$.
   (d) Conclude that P_5 is constructible.
3. Constructibility of a regular pentagon, II.
   (a) Let ξ be a primitive 5th root of 1 and $F = Q[\xi] \cap \mathcal{R}$, where $\mathcal{R}$ is the real numbers. Show that $F \neq Q$ and $F \neq Q[\xi]$.
   (b) Conclude that $\dim_Q F = 2$ and $\dim_F Q[\xi] = 2$.
   (c) Conclude that ξ, and hence P_5, is constructible.
4. Using only the constructibility of P_3 and P_5, show that P_{15} is constructible. Can you generalize your argument?
5. Let ξ be a primitive cube root of 1. Write the quadratic fields $Q[\xi]$ and $Q(P_3)$ in standard form (that is, as $Q[\sqrt{d}]$ for some square-free integer d). Verify Theorem 68 directly for $n = 3$.

6. Find a basis for $Q(P_5)$ over Q (see Problem 2; don't forget about sin 72°).

7. Show that P_n is a group under multiplication. Show that $\alpha \in P_n$ is primitive if and only if the order of α is n, and hence if $\xi \in P_n$ is primitive, then ξ^j is primitive if and only if $(j, n) = 1$.

8. Show that $x^2 + x + 1$ is irreducible as an element of $Q[x]$ but not as an element of $Z_3[x]$.

9. Let $G(x) = x^2 + x + 2 \in Z_3[x]$. Verify, by calculating, that $G(x^3) = G(x)^3$.

10. Show that if $2^t + 1$ is a prime, then $t = 2^n$ for some non-negative integer n (if k is odd, then $a^k + 1 = (a + 1)\cdot(a^{k-1} - a^{k-2} + \cdots - a + 1)$; if t is not a power of 2, then $2^t + 1$ can be written in this form). The number $2^{2^n} + 1$ is often denoted by F_n, for Fermat, who conjectured that F_n was prime for every n. $F_0 = 3$, $F_1 = 5$, $F_2 = 17$, $F_3 = 257$, and $F_4 = 65537$ are all primes; however, it has been found that F_5 through F_{16}, and a number of others, are not prime. Whether or not F_n is prime for *any* $n > 4$ is unknown.

10.5 THE GALOIS GROUP

We have seen that if the regular n-gon P_n is constructible, then $\phi(n)$ is a power of 2. In this section, we shall prove that, conversely, if $\phi(n)$ is a power of 2, then P_n is constructible. The key concept will be that of the *Galois group* of an algebraic number field.

DEFINITION. Let F be an algebraic number field. Then the *Galois group* of F, denoted by $G(F)$, is the set of all (ring) isomorphisms from F to itself.

An isomorphism from something to itself is called an *automorphism*; thus, $G(F)$ is the set of automorphisms of F. If $\sigma \in G(F)$, then $\sigma(q) = q$ for every $q \in Q$; if you don't care to prove this (see Problem 2 of section 8.2), then you can add this property to the definition of $G(F)$. If $\sigma, \tau \in G(F)$, then the product $\sigma\tau$ is defined by $\sigma\tau(\alpha) = \sigma(\tau(\alpha))$; that is, apply τ and then apply σ. It is readily verified that, with this product, $G(F)$ is a group, although it is not necessarily abelian.

THEOREM 70. Let α be an algebraic number, f its minimal polynomial, and $F = Q[\alpha]$. If $\sigma \in G(F)$, then $\sigma(\alpha)$ is a root of f. Conversely, if $\beta \in F$ is a root of f, then there is a unique $\sigma \in G(F)$ such that $\sigma(\alpha) = \beta$.

Proof: Apply σ to both sides of the equation $f(\alpha) = 0$. Remembering that σ is an automorphism, and $\sigma(q) = q$ for every $q \in Q$, we get $f(\sigma(\alpha)) = \sigma(f(\alpha)) = \sigma(0) = 0$, and so $\sigma(\alpha)$ is a root of f. Now suppose $\beta \in F$ is a root of f. The elements of F are those numbers that can be written as $g(\alpha)$ for some (nonunique) $g \in Q[x]$. Define $\sigma(g(\alpha)) = g(\beta)$. This definition depends only on the element $g(\alpha)$ and not on the particular polynomial g, for if $g(\alpha) = h(\alpha)$, then α is a root of $g(x) - h(x)$, and hence, by Corollary 2 to Theorem 41, β is a root of $g(x) - h(x)$—that is, $g(\beta) = h(\beta)$. Similarly, if $g(\beta) = h(\beta)$, then $g(\alpha) = h(\alpha)$, so σ is one-to-one. If we let $g(x) = x$, we see that $\sigma(\alpha) = \beta$.

It is easy to check that $\sigma(\theta_1 + \theta_2) = \sigma(\theta_1) + \sigma(\theta_2)$ and $\sigma(\theta_1\theta_2) = \sigma(\theta_1)\sigma(\theta_2)$ for $\theta_1, \theta_2 \in F$. To prove that σ is onto, we note that $Q[\beta]$ consists of the elements of the form $g(\beta)$ for $g \in Q[x]$; we must show that $Q[\beta] = Q[\alpha]$. Since $\beta \in Q[\alpha]$, we have $Q[\beta] \subseteq Q[\alpha]$. But α and β have the same degree, as they have the same minimal polynomial. Hence, $\dim_Q Q[\beta] = \dim_Q Q[\alpha]$, so $Q[\beta] = Q[\alpha]$ (see Problem 8 of section 7.3). Finally, to show that σ is unique, suppose $\tau(\alpha) = \beta$ for some $\tau \in G(F)$. If $g \in Q[x]$, then $\tau(g(\alpha)) = g(\tau(\alpha)) = g(\beta)$. So $\tau(g(\alpha)) = \sigma(g(\alpha))$ for every $g \in Q[x]$, and hence $\tau = \sigma$.

COROLLARY 1. If F is an algebraic number field, and $\dim_Q F = n$, then $|G(F)| \leq n$.

Proof: By Corollary 2 to Theorem 46, $F = Q[\alpha]$ for some algebraic number α. By Theorem 40, the degree of α is n. Thus, the minimal polynomial of α has at most n roots in F, so, by Theorem 70, there are at most n elements in $G(F)$.

COROLLARY 2. If F is a quadratic field, then $G(F)$ is a cyclic group of order 2.

Proof: Let $F = Q[\sqrt{d}]$, where d is a square-free integer. The minimal polynomial of $\sqrt{d}$ is $x^2 - d$, which has two roots in F: $\sqrt{d}$ and $-\sqrt{d}$. By Theorem 70, there are two elements in $G(F)$: one taking $\sqrt{d}$ to $\sqrt{d}$, and one taking $\sqrt{d}$ to $-\sqrt{d}$. Denote the first by 1 and the second by σ, so $1(a + b\sqrt{d}) = a + b\sqrt{d}$ and $\sigma(a + b\sqrt{d}) = a - b\sqrt{d}$. Clearly, $1 \cdot 1 = \sigma \cdot \sigma = 1$ and $1 \cdot \sigma = \sigma \cdot 1 = \sigma$, so $G(F)$ is a cyclic group of order 2 with generator σ.

COROLLARY 3. If ξ is a primitive nth root of 1 and $F = Q[\xi]$ then $G(F) \cong U_n$.

Proof: The minimal polynomial of ξ has $\phi(n)$ roots: the primitive nth roots of 1. These roots are the numbers of the form ξ^j, where $(j, n) = 1$, so they are all in F. Denote by $\bar{j}$ the residue class of the integer j modulo n. If $\bar{j} \in U_n$, and hence $(j, n) = 1$, let $\sigma_{\bar{j}} \in G(F)$ be the unique automorphism of F such that $\sigma_{\bar{j}}(\xi) = \xi^j$, as guaranteed by Theorem 70. This definition depends only on $\bar{j}$, and not on j itself, for if $\bar{j} = \bar{i}$ then $j - i = kn$, so $\sigma_{\bar{j}}(\xi) = \xi^j = \xi^{i+kn} = \xi^i \xi^{nk} = \xi^i = \sigma_{\bar{i}}(\xi)$. Let $f(\bar{j}) = \sigma_{\bar{j}}$; we shall show that f is an isomorphism from U_n to $G(F)$.

If $\sigma_{\bar{j}} = \sigma_{\bar{i}}$ then $\xi^j = \xi^i$, so $\xi^{j-i} = 1$. Since ξ is a primitive nth root of 1, this implies that n divides $j - i$ and hence $\bar{j} = \bar{i}$. Thus f is one-to-one. That f is onto follows from Theorem 70. Finally,

$$\sigma_{\bar{i}\bar{j}}(\xi) = \xi^{ij} = (\xi^j)^i = \sigma_{\bar{i}}(\sigma_{\bar{j}}(\xi)) = \sigma_{\bar{i}}\sigma_{\bar{j}}(\xi),$$

so $\sigma_{\bar{i}\bar{j}} = \sigma_{\bar{i}}\sigma_{\bar{j}}$ and we are done.

The important thing about Corollary 3, for our purposes, is that $G(F)$ is abelian and $|G(F)| = \dim_Q F$.

THEOREM 71. Let F be an algebraic number field such that $|G(F)| = \dim_Q F = 2^m$, and $G(F)$ is abelian. Then every element of F is constructible (given 0 and 1).

Proof: Suppose not. Let m be the least positive integer for which the theorem fails. Note that the theorem is certainly true if $m = 0$, for then $\dim_Q F = 1$ and so $F = Q$. Choose an element $\tau \in G(F)$ which is different from 1. Since $|G(F)| = 2^m$, the order of τ divides 2^m, so it is equal to 2^t for some positive integer t. Then $\sigma = \tau^{2^{t-1}}$ has order 2—that is, $\sigma \neq 1$ and $\sigma^2 = 1$.

Let $F_0 = \{\alpha \in F \mid \sigma(\alpha) = \alpha\}$. Note that F_0 is a field containing Q. Moreover $F_0 \neq F$, for if $F_0 = F$, then $\sigma = 1$. By Corollary 1 to Theorem 70, and the fact that $(\dim_Q F_0)(\dim_{F0} F) = \dim_Q F = 2^m$, we have $|G(F_0)| \dim_q F_0 \leq 2^{m-1}$. Hence, if we can find an abelian group $H \subseteq G(F_0)$ such that $|H| = 2^{m-1}$, we will have that $H = G(F_0)$ is abelian and $|G(F_0)| = \dim_Q F_0 = 2^{m-1}$. Then, by the minimality of m, we will have shown that every element of F_0 is constructible.

How do we get hold of elements in $G(F_0)$? If $\rho \in G(F)$ and $\alpha \in F_0$, then $\sigma(\rho(\alpha)) = \sigma\rho(\alpha) = \rho\sigma(\alpha) = \rho(\sigma(\alpha)) = \rho(\alpha)$, so $\rho(\alpha) \in F_0$ (here we have used the fact that $G(F)$ is abelian). Hence, ρ induces a function ρ_0 from F_0 to itself, defined, for $\alpha \in F_0$, by $\rho_0(\alpha) = \rho(\alpha)$. It is not hard to check that $\rho_0 \in G(F_0)$. Let $H = \{\rho_0 \mid \rho \in G(F)\}$. Then $H \subseteq G(F_0)$ is an abelian group; indeed $\rho_0\tau_0 = (\rho\tau)_0 = (\tau\rho)_0 = \tau_0\rho_0$, and $\rho_0{}^{-1} = (\rho^{-1})_0$. We shall show that $|H| = 2^{m-1}$.

Let $F = Q[\theta]$, $\alpha = (\theta + \sigma(\theta))/2$ and $\beta = (\theta + \sigma(\theta))/2$. Then $\theta = \alpha + \beta$, $\sigma(\alpha) = \alpha$ and $\sigma(\beta) = -\beta$. Note that $\alpha \in F_0$ and, since $\sigma(\beta^2) = \sigma(\beta)^2 = (-\beta)^2 = \beta^2$, $\beta^2 \in F_0$. Now if $\tau_0 = \rho_0$ for $\tau, \rho \in G(F)$, then $\tau(\alpha) = \tau_0(\alpha) = \rho_0(\alpha) = \rho(\alpha)$ and $\tau(\beta)^2 = \tau(\beta^2) = \tau_0(\beta^2) = \rho_0(\beta^2) = \rho(\beta^2) = \rho(\beta)^2$. Hence, $\tau(\beta) = \pm\rho(\beta)$. If $\tau(\beta) = \rho(\beta)$, then $\tau(\theta) = \rho(\theta)$, so, by Theorem 70, $\tau = \rho$; if $\tau(\beta) = -\rho(\beta)$, then $\tau(\beta) = \rho(-\beta) = \rho\sigma(\beta)$, and since $\tau(\alpha) = \rho(\alpha) = \rho\sigma(\alpha)$, we have $\tau(\theta) = \rho\sigma(\theta)$ and hence, by Theorem 70, $\tau = \rho\sigma$. Thus, if $\tau_0 = \rho_0$, then either $\tau = \rho$ or $\tau = \rho\sigma$; conversely, since σ is harmless on F_0, if $\tau = \rho$ or $\tau = \rho\sigma$, then $\tau_0 = \rho_0$. Thus, every element of $H = \{\rho_0 \mid \rho \in G(F)\}$ comes from precisely 2 elements of $G(F)$, and hence $|H| = |G(F)|/ = 2 2^{m-1}$.

Finally, since $\dim_Q F_0 = 2^{m-1}$, we have $\dim_{F0} F = 2$. Thus, $F = F_0[\theta]$ where θ satisfies a quadratic polynomial with coefficients in F_0. By the quadratic formula and Theorem 66, every element of F is constructible from elements of F_0. But the elements of F_0 are constructible, so the elements of F are constructible.

COROLLARY. A regular n-gon is constructible if and only if $\phi(n)$ is a power of 2.

Proof: The "only if" part is Corollary 1 to Theorem 69. The "if" part follows from Theorem 71 and Corollary 3 to Theorem 70.

As was mentioned in section 10.4, the ancient Greeks knew how to construct regular 3-gons and 5-gons ($\phi(3) = 2, \phi(5) = 2^2$). There was no further progress, however, until Gauss, at the beginning of the 19th century, showed that you could construct a regular 17-gon ($\phi(17) = 2^4$). His proof generalized easily to any Fermat prime. The whole thing may strike you as a mere curiosity, at best. However, this simple geometric problem has led us to examine two of the central ideas of algebraic number theory: fields generated by roots of 1 (called *cyclotomic fields*, which translates as *circle-dividing* fields, a name derived from this very problem: dividing a circle into n equal parts) and Galois groups.

10.5 PROBLEMS

1. Show that if F is an algebraic number field, then $G(F)$ is a group.
2. If $K \subseteq F$ are fields, then the *Galois group of F over K*, denoted by $G(F/K)$, is the set of (ring) isomorphisms σ of F to itself such that $\sigma(\alpha) = \alpha$ for every $\alpha \in K$.
   (a) Show that $G(F/K)$ is a group and, if F is an algebraic number field, then $G(F/Q) = G(F)$.
   (b) Generalize Theorem 70 to the case where α is algebraic over a field K, and $F = K[\alpha]$, replacing $G(F)$ by $G(F/K)$.
3. Let α be the real cube root of 2 and $F = Q[\alpha]$. Show that $|G(F)| \neq \dim_Q F$.
4. Let ω be a primitive cube root of 1 and α the real cube root of 2. Let $F = Q[\omega, \alpha]$.
   (a) Show that $x^3 - 2$ is the minimal polynomial of α over $Q[\omega]$ and $x^2 + x + 1$ is the minimal polynomial of ω

over $Q[\alpha]$. (Use the fact that $\dim_Q F = 3 \dim_{Q[\alpha]} F = 2 \dim_{Q[\omega]} F$.)

(b) Using part (b) of Problem 2, show that there is a $\sigma \in G(F/Q[\omega])$ such that $\sigma(\alpha) = \omega\alpha$, and a $\tau \in G(F/Q[\alpha])$ such that $\tau(\omega) = \omega^2$.

(c) Conclude that there exist $\sigma, \tau \in G(F)$ such that $\sigma(\omega) = \omega$, $\sigma(\alpha) = \omega\alpha$, $\tau(\omega) = \omega^2$, and $\tau(\alpha) = \alpha$. Show that $\sigma\tau \neq \tau\sigma$, so $G(F)$ is not abelian.

(d) Show that $|G(F)| = \dim_Q F = 6$.

5. Show that if G is an abelian group of order p^m, p a prime number, then G has an element of order p.

6. Let ξ be a primitive 5th root of 1 and $F = Q[\xi]$. Find a $\sigma \in G(F)$ such that σ has order 2. What is $F_0 = \{\alpha \in F \mid \sigma(\alpha) = \alpha\}$? What is $G(F_0)$?

INDEX